ARTIFICIAL INTELLIGENCE IN AGRICULTURE 2001

(AIA 2001)

A Proceedings volume from the 4th IFAC/CIGR Workshop
Budapest, Hungary, 6 - 8 June 2001

Edited by

I. FARKAS

Department of Physics and Process Control,
Szent István University, Gödöllő, Hungary

Published for the

INTERNATIONAL FEDERATION OF AUTOMATIC CONTROL

by

PERGAMON

An Imprint of Elsevier Science

UK	Elsevier Science Ltd, The Boulevard, Langford Lane, Kidlington, Oxford, OX5 1GB, UK
USA	Elsevier Science Inc., 660 White Plains Road, Tarrytown, New York 10591-5153, USA
JAPAN	Elsevier Science Japan, Tsunashima Building Annex, 3-20-12 Yushima, Bunkyo-ku, Tokyo 113, Japan

First edition 2001

Library of Congress Cataloging in Publication Data

A catalogue record for this book is available from the Library of Congress

British Library Cataloguing in Publication Data

A catalogue record for this book is available from the British Library

ISBN 0-08-043563 7
ISSN 1474-6670

Transferred to digital printing 2005

4th IFAC/CIGR WORKSHOP ON ARTIFICIAL INTELLIGENCE IN AGRICULTURE 2001

Sponsored by
International Federation of Automatic Control (IFAC)
IFAC Technical Committees on
- Modelling and Control in Agricultural Processes
- Control Systems in Agriculture

Co-Sponsored by
International Commission of Agricultural Engineering (CIGR)

Organized by
Department of Physics and Process Control, Szent István University, Gödöllő, Hungary
Hungarian Academy of Sciences - SZIUG, Process Control Research Group

International Programme Committee (IPC)
Farkas, I. (H) (Chairman)
Murase, H. (J) (Vice-Chairman)
Xiong, F.L. (PRC) (Vice-Chairman)

Hasimoto, Y. (J)
De Baerdemaeker, J. (B)
Bot, G. (NL)
Day, W. (UK)
Dijkhunisen, A.A. (NL)
Gauthier, L. (CDN)
Grenier, P. (F)
Gu, S.G. (PRC)
Hirafuji, M. (J)
Hoshi, T. (J)
Jacucci, G. (I)
Martin-Clouaire, R. (F)
Morimoto, T. (J)
Munack, A. (D)
Nakamura, K. (J)
Noguchi, N. (J)
Nybrant, T. (S)
Rafea, A. (ET)
Satake, T. (J)
Schiefer, G. (G)
Seginer, I. (IL)
Sevila, F. (F)
Sigrimis, N.A. (GR)
van Straten, G. (NL)
Tantau, H.-J. (D)
Toyoda, K. (J)
Udink ten Cate, A.I. (NL)
Wagner, P. (G)
Young, P.C. (UK)

National Organizing Committee (NOC)
Buzás, J. (Chairman)

Farkas, I. (Convenor)
Fekete, A.
Lágymányosi, A.
Neményi, M.
Szűcs, M.
Tóth, M

PREFACE

The scope of the AIA'2001 Workshop was to provide a forum for the presentation of new results in research, development and applications of Artificial Intelligence in Agriculture (AIA). The Workshop aimed at bringing together leading researchers and practitioners from academy and industry in order to favour the emergence or consolidation of bridges between AI and its applications in Agriculture and domains connected to it (in particular, environmental sciences).

The presented papers were selected into five sessions as Fundamentals; Decision tools; Neural networks; Quality and measurements; Fuzzy modelling and control and Modelling and control.

The IFAC Workshop was sponsored by IFAC Technical Committees on "Modelling and Control in Agricultural Processes" and "Control Systems in Agriculture". The main international co-sponsor was the International Commission of Agricultural Engineering (CIGR). The additional local sponsors were the Szent István University Gödöllő, Hungary, the Hungarian Academy of Sciences and the Ministry of Education, Hungary. The Workshop was locally organised by the Department of Physics and Process Control, Szent István University Gödöllő, Hungary and the Process Control Research Group, Hungarian Academy of Sciences. The venue of the Workshop was partly the HotelÉben in Budapest and the Campus of the Szent István University Gödöllő.

Prof. I. Farkas
Editor

CONTENTS

FUNDAMENTALS, DECISION TOOLS

NEURAL NETWORKS

QUALITY AND MEASUREMENTS

FUZZY MODELLING AND CONTROL

MODELLING AND CONTROL

PSEUDO-DERIVATIVE FEEDBACK BASED IDENTIFICATION OF UNSTABLE PROCESSES WITH APPLICATION TO A BIOREACTOR

K.G.Arvanitis[1], G.D.Pasgianos[2], E.Kyriannakis[2] and N.A.Sigrimis[1]

[1] *Agricultural University of Athens, Dept. of Agricultural Engineering, Iera Odos 75, 11855, Athens, GREECE, email: karvan@aua.gr, n.sigrimis@computer.org*
[2] *National Technical University of Athens, Dept. of Electrical and Computer Engineering, Zographou 15773, Athens, GREECE, email: pasgianos@24hteacher.com, ekyrian@robotics.ece.ntua.gr*

Abstract: The use of a pseudo-derivative feedback structure for parameter identification of unstable first order plus dead-time processes is investigated. Two alternative identification methods are presented. The proposed identification methods require small computational effort and they are particularly useful for on-line applications. They ensure considerably smoother response (i.e. less overshoot), than known identification methods based on standard PI/PID controllers, while providing better accuracy and more simplicity in implementation, due to the simpler structure of the pseudo-derivative feedback controller used. *Copyright © 2001 IFAC*

Keywords: Pseudo-derivative feedback, identification, process control, unstable models.

1. INTRODUCTION

Research on tuning methods of conventional controllers for unstable first order processes with dead-time (UFOPDT) has recently been very active (see, Srinivas and Chidambaram 1996; Chidambaram, 1997, and the references cited therein). The common feature of the existing tuning methods for UFOPDT systems is that they give excessive overshoot. Jacob and Chidambaram (1996) were the first to point out this drawback and proposed a new tuning rule, which incorporates both a two-stage P-PI control structure and the internal model controller (IMC) tuning rule (Morari and Zafiriou, 1986). All these simple controller-tuning methods mentioned above are applicable as far as the parameters of the unstable process under control are available, e.g. through an identification procedure. Attempts to extend to the unstable system case, known identification methods for stable processes with time-delay, such as the well-known Yuwana and Seborg method (Yuwana and Seborg, 1982), have been reported in the literature (Kavdia and Chidambaram, 1996). In particular, this method first identifies a second order plus time delay model, for the closed-loop system using the response obtained for a set-point step change of a proportional controller. Then, from the identified closed-loop model, the open-loop transfer function is derived. Attempts to improve the method in Kavdia and Chidambaram (1996) have also been reported in Srinivas and Chidambaram (1996). However, for UFOPDT systems a large offset is obtained by using a proportional-stabilizing controller (Venkatashankar and Chidambaram, 1994). In real situations, such a test is not allowed. To avoid this drawback, in Ananth and Chidambaram (1999) another extension of the Yuwana and Seborg method, has been proposed. This method relies on using a PI stabilizing controller instead of the P controller proposed in previous methods, in order to avoid large offsets. However, as mentioned above, PI controllers produce large overshoots in the case of UFOPDT processes. For this reason, in Ananth and Chidambaram (1999), the use of a PID stabilizing controller has been suggested instead of the PI controller, since a PID controller significantly reduces the overshoot and obviously gives zero offset. However, in this method a three-parameter controller needs to be tuned.

This paper investigates some aspects of the configuration developed by Phelan (1978), and called the "Pseudo-derivative Feedback Controller" (PDF), when applying it to the identification of UFOPDT processes. The PDF controller is a diversity of the conventional PID controller that differs from it at the following main points: (a) In contrast to the standard PID controller configuration, wherein all control elements are located to the forward path, the elements of the PDF controller are appropriately located in both the forward and the feedback path. This configuration contributes to better understanding the controller action, since the elements located to the feedback path are mainly dedicated in assigning the desired closed-loop performance, while the forward path elements are devoted to steady state error elimination. (b) The conventional PID controller acts on the process error with the result that, each element contributes to closed-loop poles as well as zeros. In contrast, the PDF controller elements do not contribute to closed-loop zeros, and hence, do not deteriorate any overshoot of the closed-loop response.

In this paper, we apply to UFOPDT processes a simple case of the general PDF control scheme, containing only proportional action in the feedback path and called the PD-0F control. PD-0F controllers have successfully been used by Arvanitis *et al.* (2000), in order to control unstable processes. In the present paper, the possibility of using the PD-0F controller for identification of unknown UFOPDT processes is investigated. Two alternative methods for identification of UFOPDT models are presented in the paper. Both methods are similar to the method reported in Ananth and Chidambaram (1999), with the difference that a PD-0F controller is used here to stabilize the system, instead of a PI or a PID controller. The reason for using this alternative scheme is due to the inherent feature of the PD-0F controller to produce no excessive overshoot, a fact that renders the derivative action of the PID controller, used in Ananth and Chidambaram (1999), unnecessary. The first method identifies the gain and the time constant of the process from the closed-loop step response, while the dead time is obtained from the initial portion of this response. It requires small computational burden and is particularly useful in cases where the dead time of the process is easily measurable (e.g. large dead time, absence of measurement noise, etc.). The second method identifies all three parame-ters of the UFOPDT model, from the closed-loop step response, using the Newton-Raphson's techni-que to solve a set of nonlinear equations, involving the model parameters sought. The proposed identification methods require small computational effort and they are particularly useful for on-line applications. It is mentioned at this point that, when applying the above identification methods, the controller settings are calculated through the PD-0F tuning rules reported in Arvanitis *et al* (2000). Finally, an application of the proposed methods to the open loop unstable transfer function model describing a biological reactor with hard input constraints and significant measurement delay is also presented. The obtained results reveal that the proposed methods provide considerably smoother response than known identification methods based on standard PI/PID controllers, while providing better accuracy and more simplicity in implementation, due to the simpler structure of the two parameter PD-0F controller.

2. THE PD-0F CONTROL STRUCTURE

The general PDF control is a modification of the integral control with derivative-feedback algorithm (IDF). The IDF algorithm has an integrator in the forward path where the controller is usually located and derivative feedback on the controlled variable. Mathematically, this idealized controller works well, but in practical applications, derivatives of the control variable show significant noise, especially for the second and higher order derivatives. The general PDF control is different. Instead of feeding back the derivative of the controlled variable, in order to calculate the actuating signal, then integrating this component in the forward loop, all of the controlled variable derivative feedback is bypassed to the integrator output. This approach reduces the number of derivatives by one. It is worth noticing at this point that, although the control coefficients for the IDF algorithm and the general PDF control structure are different, simulation show that they have the same response.

We next focus our attention to the simplest case of the general PDF control structure, depicted in Fig. 1 and called the PD-0F control structure. We shall next analyze the behavior of this specific feedback scheme, especially in the case where the system under control is an UFOPDT process with transfer function model of the form

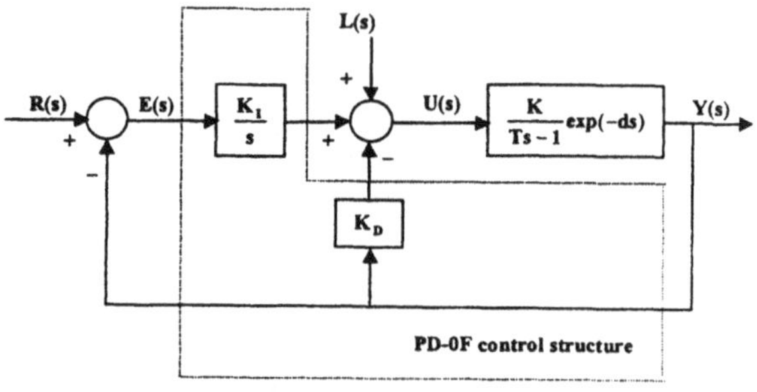

Fig. 1. The PD-0F control structure applied to an UFOPDT process model.

$$G_P(s) = K\exp(-ds)/(Ts-1) \qquad (1)$$

where K, d and T are the process gain, the time delay and the time constant, respectively. To this end, observe that the transfer function of the closed-loop system $G_{CL}(s)$ takes the form

$$G_{CL}(s) = K_I G_p(s)/[s + (K_D s + K_I)G_p(s)]$$

which, for the case of UFOPDT models of the form (1) takes the special form

$$G_{CL}(s) = \exp(-ds)/\left\{s\left[T(KK_I)^{-1}s - (KK_I)^{-1}\right] + \left(K_D K_I^{-1}s + 1\right)\exp(-ds)\right\} \quad (2)$$

Now, using the approximation

$$\exp(-ds) \cong 1 - ds$$

in the denominator of (2), we obtain

$$G_{CL}(s) = \frac{\exp(-ds)}{\tau_e^2 s^2 + 2\zeta\tau_e s + 1} \quad (3)$$

i.e. a second order system with dead time, where

$$\tau_e = \sqrt{\frac{K_D}{K_I}\left(\frac{T}{KK_D} - d\right)}, \zeta = \frac{\frac{K_D}{K_I}\left(1 - \frac{1}{KK_D}\right) - d}{2\sqrt{\frac{K_D}{K_I}\left(\frac{T}{KK_D} - d\right)}} \quad (4)$$

3. IDENTIFICATION OF UFOPDT MODELS USING PD-0F CONTROLLERS.

In this Section, two new closed-loop identification methods for UFOPDT models will be presented. These methods are extensions of the Yuwana and Seborg method, in the case of UFOPDT processes. Both methods use a PD-0F controller in order to stabilize the process. The proposed methods provide all the benefits of the PID-based identification method reported in Ananth and Chidambaram (1999), and require less effort, since it is simpler to seek for a two-term stabilizing controller. The first method identifies the gain and the time constant of the process from the closed-loop step response, while the dead time is obtained from the initial portion of this response. The second method identifies all three parameters of the UFOPDT model, from the closed-loop step response, using the Newton-Raphson's technique in order to solve a set of nonlinear equations, involving the model parame-ters sought. Both methods require small computational effort and are particularly useful in cases where the dead time of the process is easily measurable.

In view of relations (3), (4), the three fundamental steps of the first identification method proposed here, are similar to those of the method reported in Ananth and Chidambaram (1999). These steps are summarized as follows:

Identification method I:

Step 1. The dead time, d, of the process is obtained from the initial portion of the step response.

Step 2. For a given set point step change, the closed loop response is obtained as in Fig. 2. Let now y_{p_1} and y_{p_2} be the first and second peak of the output response, respectively, y_{m_1} be the first minimum of the output response, y_∞ be the steady state value of the output response, and Δt be the period of the oscillation. From the values of the measurable quantities of y_{p_1}, y_{p_2}, y_{m_1}, y_∞ and Δt the values of τ_e and ζ can be calculated according to the following equations (see Yuwana and Seborg (1982) for its derivation)

$$\tau_e = \frac{\Delta t}{2\pi}\sqrt{1-\zeta^2}\ ,\ \zeta = \frac{\zeta_1 + \zeta_2}{2} \quad (5)$$

where

$$\zeta_1 = \frac{-\ln\left(\frac{y_\infty - y_{m_2}}{y_{p_1} - y_\infty}\right)}{\sqrt{\pi^2 + \left[\ln\left(\frac{y_\infty - y_{m_2}}{y_{p_1} - y_\infty}\right)\right]^2}}$$

$$\zeta_2 = \frac{-\ln\left(\frac{y_{p_2} - y_\infty}{y_{p_1} - y_\infty}\right)}{\sqrt{4\pi^2 + \left[\ln\left(\frac{y_{p_2} - y_\infty}{y_{p_1} - y_\infty}\right)\right]^2}}$$

From the values of ζ and τ_e obtained from equations (5), the dominant closed loop poles $\sigma \pm j\omega$ can be obtained as follows

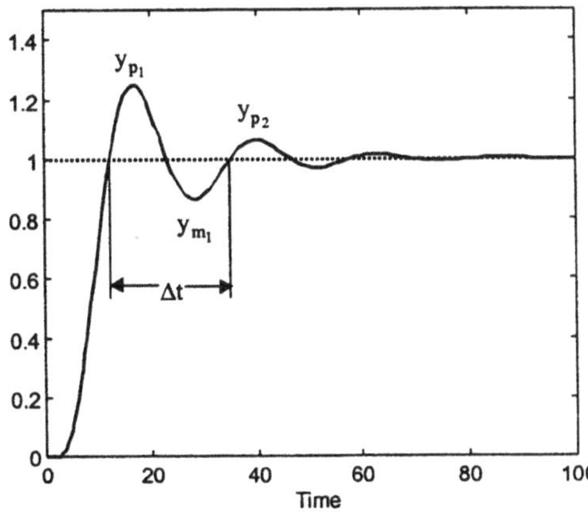

Fig. 2. Closed-loop response of an UFOPDT model controlled using the PD-0F control structure.

$$\sigma = -\zeta/\tau_e\ ,\ \omega = \sqrt{1-\zeta^2}/\tau_e \quad (6)$$

Step 3. Substituting the values of the dominant poles of the closed-loop system in the denominator of (2) and equating the real and imaginary parts of the resulting equation to zero, we obtain the following two algebraic equations with respect to K and T

$$e^{-d\sigma}\left[K_D\sigma\cos(d\omega) + K_D\omega\sin(d\omega) + K_I\cos(d\omega)\right]K + \left(\sigma^2 - \omega^2\right)T = \sigma \quad (7a)$$

$$e^{-d\sigma}\left[-K_D\sigma\sin(d\omega)+K_D\omega\cos(d\omega)-K_I\sin(d\omega)\right]K + 2\sigma\omega T = \omega \quad (7b)$$

Defining

$$A = e^{-d\sigma}\left[K_D\sigma\cos(d\omega)+K_D\omega\sin(d\omega)+K_I\cos(d\omega)\right] \quad (8a)$$
$$B = \sigma^2-\omega^2$$
$$C = e^{-d\sigma}\left[-K_D\sigma\sin(d\omega)+K_D\omega\cos(d\omega)-K_I\sin(d\omega)\right] \quad (8b)$$
$$E = 2\sigma\omega$$

and solving (7a) and (7b) with respect to K and T we finally obtain

$$K=\frac{E\sigma-B\omega}{AE-BC}, \quad T=\frac{A\omega-C\sigma}{AE-BC} \quad (9)$$

This completes the method.

It is worth noticing at this point that, for the identification of the model parameters using the methods reported in Yuwana and Seborg (1982) and Kavdia and Chidambaram (1996), the parameter K is obtained from the steady state offset. After obtaining K, the values of T and d are calculated from the characteristic equations using the dominant poles (Kavdia and Chidambaram, 1996). However, it is not difficult to see that, in the case where a PD-0F controller is used to identify the model, there is no steady-state offset. For this reason, similarly to the PID based identification method reported in Ananth and Chidambaram (1999), the value of d is first obtained from the initial portion of the closed-loop response, and then (7a) and (7b) are used, in order to obtain K, T.

The second PD-0F controller based identification method is as follows:

Identification Method II:

Step 1. For a given set point change, the gain of the process is obtained through the relation

$$K=-\frac{y_\infty}{u_\infty} \quad (10)$$

where, u_∞ is the steady state value of the stabilizing PD-0F controller output.

Step 2. The second step of the Identification Method II coincides with Step 2 of Identification Method I.
Step 3. Using an argument similar to that reported in the third step of the Identification Method I, form equations (7a) and (7b). Then, solving (7a) with respect to T, we obtain

$$T=\frac{\sigma-e^{-d\sigma}\left[K_D\sigma\cos(d\omega)+K_D\omega\sin(d\omega)+K_I\cos(d\omega)\right]K}{\sigma^2-\omega^2} \quad (11)$$

where, K is obtained by (10). Substituting (11) into (7b), yields

$$e^{-d\sigma}\left[-K_D\sigma\sin(d\omega)+K_D\omega\cos(d\omega)-K_I\sin(d\omega)\right]K + 2\sigma\omega\times\left\{\frac{\sigma-e^{-d\sigma}\left[K_D\sigma\cos(d\omega)+K_D\omega\sin(d\omega)+K_I\cos(d\omega)\right]K}{\sigma^2-\omega^2}\right\} = \omega \quad (12)$$

Simple algebraic manipulations in (12) yield the following equation

$$L_1\sin(d\omega)+L_2\cos(d\omega)+L_3\exp(d\sigma)=0 \quad (13)$$

where

$$L_1=-K_D\sigma\left(\frac{\sigma^2+\omega^2}{\sigma^2-\omega^2}\right)-K_I$$
$$L_2=-K_D\omega\left(\frac{\sigma^2+\omega^2}{\sigma^2-\omega^2}\right)-K_I\left(\frac{2\sigma\omega}{\sigma^2-\omega^2}\right)$$
$$L_3=\frac{\omega}{K}\left(\frac{\sigma^2+\omega^2}{\sigma^2-\omega^2}\right)$$

Obviously, equation (13) is nonlinear and has no direct solution. However, if a close initial guess is available, the Newton-Raphson's method can be used to obtain a sufficiently accurate solution after a few iterations. To derive a good initial guess, we start by approximating the sine, cosine and expo-nential function by the following second-order poly-nomials (for details, see Wang *et al.* (1999))

$$\sin(x)=\mu x^2+\nu x, \quad \cos(x)=\mu x^2+\xi x+1 \quad (14a)$$
$$\exp(x)=0.5x^2+x+1 \quad (14b)$$

where,

$$\mu=\frac{8(1-\sqrt{2})}{\pi^2}, \quad \nu=\frac{2(2\sqrt{2}-1)}{\pi}, \quad \xi=\frac{2(2\sqrt{2}-3)}{\pi}$$

Note that the fittings for the sine and the cosine function are exact at the points x=0, π/4, π/2. Applying (14a) and (14b) to (13), we obtain

$$\left(L_1\mu\omega^2+L_2\mu\omega^2+L_3\frac{\sigma^2}{2}\right)d^2 + (L_1\nu\omega+L_2\xi\omega+L_3\sigma)d+(L_2+L_3)=0 \quad (15)$$

Solving (15) and taking the smaller absolute root yields the first approximate d_1. With this initial value, the Newton-Raphson's method is applied in the usual way

$$d_2=d_1-\frac{\phi(d_1)}{\phi^{(1)}(d_1)} \quad (16)$$

where

$$\phi(d)=L_1\sin(d\omega)+L_2\cos(d\omega)+L_3\exp(d\sigma)$$
$$\phi^{(1)}(d)=\omega L_1\cos(d\omega)-\omega L_2\sin(d\omega)+\sigma L_3\exp(d\sigma)$$

Simulation results show that the approximate provided by (16) is remarkably close to the true value and a 95% accuracy can usually be achieved, provided that the dead time is not exceptionally large. Hence, only one iteration is usually needed in the Newton-Raphson's method for fine identification of the time delay.

When, d is obtained through the above procedure, the process time constant is identified through (4.10). This completes the method.

4. SIMULATION STUDIES

4.1. Numerical Examples

In order to demonstrate the effectiveness of the proposed methods and to provide a comparison with the existing identification methods, a numerical example is first elaborated. In particular, the unstable process model studied in Ananth and Chidambaram (1999) is considered. Parameter values for this model are K=4, d=2 and T=4. In the case where a P controller is used to identify the process parameters, a large offset of 170% is obtained. However, in process industry such an offset will not be allowed. Therefore, in order to eliminate this offset, the use of PI or PID controller is indispensable. If we use the method reported by Ananth and Chidambaram (1999), with a PI controller having the parameters K_C=0.343, τ_I=29.412, the identified model parameters are $K=4.011$, T=4.069, d=2.0. On the other hand, by applying the method of Ananth and Chidambaram (1999), with a PID controller, whose parameters are K_C=0.250, τ_I=22.750, τ_D=1.365, the identified model parameters are K=3.977, T=3.868, d=2.0.

We next apply Identification Method I, based on the PD-0F control structure, for the purpose of identifying the model parameters. An initial set of the PD-0F based controller parameters is given by K_D= 0.3968, $K_I = 0.0325$. With these settings, the response of the closed-system is analogous to that of Fig. 2. The dead time is obtained from the initial portion of the response and has the value d=2.0. The values of $y_{p_1}, y_{m_1}, y_{p_2}$ and Δt are summarized in the first line of Table 1, together with the values of τ_e and ζ obtained by equations (5), the values of σ and ω obtained by equations (6), as well as the parameters of the PD-0F controller. The identified model parameters K and T are then obtained from equations (7)-(9) and they are also given in the first line of Table 1.

Results from the application of Identification Method I on other UFOPDT process models are also listed in Table 1. In the case where, a PD-0F control structure with K_I=K_D can be used, the results are also listed in Table 1. It is worth noticing at this point that, for the last case of Table 1, i.e. the case of the process with actual parameters K=1; T=1; d=0.8, in Ananth and Chidambaram (1999), a three-term (PID) controller needs to be used in order to perform the identification, since a PI controller cannot stabi-lize the process. However, using the proposed me-thod, identification of this process is plausible by the use of a two-term controller, namely the PD-0F controller. Applying the proposed Identification Method I, the parameters of the identified model are K= 0.9959, T=0.9854 and d=0.8, which match well with the actual process parameters.

We next apply Identification Method II, for the purpose of identifying the parameters of the UFOPDT process model, with actual parameters K=4, d=2 and T=4. In this case, an initial set of the PD-0F based controller parameters is given by K_D=0.3779, $K_I = 0.0178$. The values of $y_{p_1}, y_{m_1}, y_{p_2}$, Δt, τ_e, ζ, σ and ω are summarized in the appropriate line of Table 1. The identified model parameters are K= 4.0003, T=3.9416, d=2.0036. Results from the application of identification method II on other UFOPDT process models are also listed in the last three lines of Table 1.

4.2. Application to an unstable bioreactor

The open loop behavior of a variety of constant volume continuous stirred tank fermenters (CSTF) with sterile feed, can be described by the following unstructured model (Agarwal and Lim, 1986)

$$\frac{dX}{dt} = (\mu(S) - D)X \ , \ \frac{dS}{dt} = D(S_f - S) - \frac{\mu(S)X}{Y_{X/S}} \quad (17)$$

where, $\mu(S) = \mu_m S(K_m + S + S^2/K_I)^{-1}$ is the specific growth rate, $Y_{X/S}$ is the cell-mass yield coefficient, μ_m is the maximum specific growth rate, K_m is the growth rate constant and K_I is the substrate inhibition constant. Typical values for the model parameters are (Agarwal and Lim, 1986)

$$Y_{X/S} = 0.4\%g/g, \ S_f = 4\%g/g, \ D = 0.36h^{-1}$$

$$\mu_m = 0.53h^{-1}, \ K_m = 0.12\%g/g, \ K_I = 0.4545\%g/g$$

The solution of (17) exhibits an unstable steady sta-te at $[X,S]_2$=[0.9951, 1.5122]. In the present simula-tion study it is desired to operate the CSTF at this unstable steady state. The cell mass concentration X is the controlled variable. The upper and lower constraints to the manipulated variable D, are D_L= 0.25 h^{-1} and D^U=0.40 h^{-1}. A measurement delay of one hour is also considered in the measurement of X.

In the present simulation study, the proposed Identification Method I is used in order to obtain an unstable first-order plus dead-time model of the above continuous bioreactor. The nonlinear model equation are solved along with the PD-0F controller with $K_I = -0.1$ and $K_D = -0.5$. The closed-loop response in X, for a step change in the set point of X from 0.9951 to 1.1941 is obtained. From the response we get the peak values in the $\Delta X = X - X_S$ as $y_{p_1} = 0.2537$, $y_{m_1} = 0.1853$, $y_{p_2} = 0.2028$, respectively, while $\Delta t = 11.7594$. On the basis of equations (5), we get $\tau_e = 1.7020$ and $\zeta = 0.4160$. Moreover, $\sigma = -0.2444$ and $\omega = 0.5343$. Finally, the model parameters are obtained from equations (7)-(9). The identified model parameters obtained from the application of the proposed Identification Method I are $K = -6.0406$, $T = 5.0780$ and $d = 1.0$.

We next apply the proposed Identification Method II, to the nonlinear model of the bioreactor, in order to obtain an UFOPDT transfer function model. To this end, the nonlinear model equation are solved along with the PD-0F controller with K_I=-0.1361 and K_D= -0.5516. The closed-loop response in X, for a step change in the set point of X from 0.9951 to 1.1941 is obtained. From the response we get the peak values of $\Delta X = X - X_S$ as $y_{pl}=0.2994$, $y_{ml}=0.0952$, $y_{p_2} = 0.0389$, respectively, while $\Delta t = 11.9257$. On the basis of equations (5), we get $\tau_e = 1.7951$ and $\zeta=0.3249$. Moreover, $\sigma = -0.1810$ and $\omega = 0.5269$. Finally, the model parameters are obtained from equations (7)-(9). The identified model parameters obtained from the application of the proposed Identification Method II are $K = -5.8900$, T=5.8592 and $d = 1.0006$.

The identified transfer function model parameters, obtained from the application of both identification methods, can be compared with the parameters of the model obtained by linearization around the operating point. The linear model obtained by local linearization of the nonlinear model around the unstable steady state is

$$G_P(s) = \frac{-5.89}{5.86s - 1}\exp(-s)$$

Obviously, in both cases, the identified model parameters match satisfactorily with those of the linearized model. In particular, the results of Identification Method II are very close to those obtained by local linearization.

5. CONCLUSIONS

In this paper, the problem of identifying the parameters of UFOPDT processes has been investigated, and the use of pseudo-derivative feedback controllers has been proposed to address the problem. Two alternative identification methods have been presented, which require small computational effort and they are particularly useful for on-line applications. They also ensure considerably smoother response, than known identification methods based on standard PI/PID controllers, while providing better accuracy and more simplicity in implementation, due to the simpler structure of the pseudo-derivative feedback controller as compared to PID controllers.

REFERENCES

Agarwal P. and H.C.Lim (1986). Analysis of various control schemes for continuous bioreactors, *Adv. Biochem. Biotechnology*, **30**, 61-90.

Ananth, I. and M.Chidambaram (1999). Closed-loop identification of transfer function model for unstable systems., *J.Franklin Inst.*, **336**, 1055-1061.

Arvanitis, K.G., G.D.Pasgianos and N.A. Sigrimis (2000). An alternative feedback control structure for open-loop unstable systems. *Proc. XIV Memorial CIGR World Congress 2000*, Tsukuba, Japan, November 28-December 1, 2000, 947-952.

Chidambaram, M. (1997). Control of unstable systems: A review. *J. Energy, Heat, Mass Transf.*, **19**, 49-56.

Jacob, E.F. and M. Chidambaram (1996). Design of controllers for unstable first-order plus time delay systems. *Comp. & Chem. Eng.*, **20**, 579-584.

Kavdia, M. and M.Chidambaram (1996). On-line controller tuning for unstable systems. *Comp. & Chem. Eng.*, **20**, 301-305.

Morari, M. and E. Zafiriou (1989). *Robust Process Control.* Prentice-Hall, Englewood Cliffs, N.J.

Phelan, R.M., (1978). *Automatic Control Systems,* Cornell University Press, Ithaca, New York.

Srinivas, N. and M.Chidambaram (1996). Comparison of on-line controller tuning methods for unstable systems. *Proc. Contr. Quality*, **8**, 177-183.

Venkatashankar, V. and M.Chidambaram (1994). Design of P and PI controllers for unstable first-order plus time delay systems, *Int. J. Control,* **60**, 137-144.

Wang, Q.G., T.H.Lee, H.W.Fung, Q.Bi and Y. Zhang (1999). PID tuning for improved performance, *IEEE Trans. Control Syst. Techn.*, **7**, 457-465.

Yuwana, M. and D.E.Seborg (1982). A new method for on-line controller tuning. *AIChE Journal,* **28**, 434-440.

System			Controller Settings		Process Characteristics								Identified model			Identification Method
K	T	d	K_p	K_I	y_{p_1}	y_{m_1}	y_{p_2}	Δt	τ_e	ζ	σ	ω	K	T	d	
4	4	2.0	0.3968	0.0325	1.6546	0.3898	1.5609	19.3594	3.0802	0.0245	-0.0080	0.3246	4.0022	3.9853	2.0	I
1	1	0.1	1.6667	1.6667	1.5180	0.7314	1.1392	4.5502	0.7089	0.2047	-0.2888	1.3809	1.0000	0.9999	0.1	I
1	1	0.1	5.0616	15.620	1.3297	0.8805	1.0433	1.1966	0.1812	0.3074	-1.6965	5.2511	0.9991	0.9991	0.1	I
3	2	0.4	1.0585	1.0585	1.7061	0.4233	1.4708	3.5848	0.5694	0.0644	-0.1131	1.7527	2.9945	1.9972	0.4	I
3	2	0.4	1.0946	0.9639	1.5659	0.6109	1.2670	3.5665	0.5636	0.1186	-0.2105	1.7617	2.9992	1.9969	0.4	I
1	1	0.8	1.1328	0.0333	1.3621	0.6042	1.3398	11.7900	1.8764	0.0095	-0.0051	0.5329	0.9959	0.9854	0.8	I
4	4	2	0.3779	0.0178	1.2526	0.8692	1.0611	22.1000	3.4342	0.2161	-0.0629	0.2843	4.0003	3.9416	2.0036	II
1	1	0.1	5.6445	21.796	1.4582	0.7584	1.1287	0.9921	0.1547	0.1989	-1.2851	6.3332	1.0000	1.0104	0.1010	II
3	2	0.4	0.9340	0.4753	1.2333	0.9357	1.0177	5.1180	0.7536	0.3796	-0.5037	1.2277	3.0000	2.0013	0.4009	II
1	1	0.8	1.1294	0.0254	1.1683	0.7657	1.1440	11.0631	1.7594	0.1	-0.01	0.5679	1.0000	0.9303	0.8026	II

Table 1. Numerical results obtained from the application of the proposed UFOPDT process identification methods on various systems.

DEVELOPING STRATEGIC EXPERT SYSTEM USING MULTIPLE DESIGN APPROACH

Soliman A. Edrees * Ibrahim Fathy* Mohamed Yahia* Ahmed Rafea **

* Central Lab for Agricultural Expert System (CLAES), El-Noor St., Dokki, Giza, EGYPT
{ **soliman, ibrahim, mdyehia** }@mail.claes.sci.eg
** Computer Science Department, American University in Cairo, rafea@aucegypt.edu

ABSTRACT: This paper presents a strategic Expert System for Wheat crop production, which had been developed using Multiple Design approach. This Expert System consists of six subsystems namely; variety selection, land preparation, sowing, irrigation, fertilization, and harvest. These subsystems exchange data between them through, a common databases. Each subsystem advises wheat growers strategic plan(s), which is a set of agricultural/chemical operations before or during growing season. Applying an appropriate strategic plan avoid farmers from having a problem during the growing season. Copyright ©2001 IFAC

Keywords : Expert System, knowledge Based System, Hierarchical System, Hierarchical Structure, Hierarchically Intelligent

1. INTRODUCTION

Expert Systems have found wide applicability in problem solving of agricultural crop management. In Egypt, the Central Laboratory of Agricultural Expert Systems (CALES) has been established. The aim of establishment this laboratory is to make use of Expert Systems technology in transferring knowledge from agricultural Domain Experts to Extension officers and farmers.

CLAES has developed so many Expert Systems in agricultural production and animal health domain. Applying these expert system in the filed gave a good impact in three dimension namely: economically, environmentally, and human resources. From economic view, applying these systems had increased the agricultural production at least 20%. In the environment impact these systems reduced the use of pesticides, chemical fertilizers, and optimize the water quantity. In human resources impact, the developed systems have increased the performance of extension officers. (Rafea 1998).

NEPER wheat expert system was one of the Expert System developed in CLAES. This system is verified, validated, and tested in the field. The system has six components. Each component represents an individual expert system, which is capable of running alone or integrated with the other components. A common database was used to share the common data. These components are variety selection, land preparation, sowing, irrigation, fertilization, and harvest subsystems. NEPER is now ready as a production product valid to work as a tool used by Extension Officers.

The Multiple Design technique (Kamel et al 1994) was used in developing NEPER strategic expert system. The need to use this technique comes from the nature of domain problem. Since the system gives strategic plans in advance for the farmers to grow wheat and the farmer wants to select a suitable plan among these plans. So we found that Multiple Design approach is suitable for developing NEPER expert system.

Section two of this paper introduces a background of routine design and multiple design. Section three describes the conceptual design of NEPER expert system. Section four describes the implementation aspects and case study. Section five is a conclusion.

2. ROUTINE DESIGN AND MULTIPLE DESIGN

Routine design approach is based on hierarchy of cooperating specialists, each responsible for identified part of a complete design. The higher level specialists in the hierarchy typically represent more conceptual aspects of the design process, whereas the lower level specialists represent more parametric aspects of the design process.

In routine design, (Tong 1991), each specialist follows one of a set of pre-specified plans. Plans prescribe the problem solving actions to be followed and are defined at the time of building a design system. While the pieces of the design process (plans) are known in advance, the combination of pieces as well as their ordering is typically determined during the actual problem solving. While each specialist has a predefined set of plans to choose from, the actual plan to be followed is dynamically chosen during problem solving on the input design requirements as well as the results of the design established so far. The choice of plans for every specialist can therefore lead to the generation of a design which is novel due to the use of plans by the different specialists that have not been used together previously.

Problem solving knowledge in Routine design is represented in the form of a collection of agents of varying types. (Kamel, et. al. 94a) identify types of agents namely: Specialist, Plan, Plan sponsor, Task, Step, Plan selector, Constraints, Table matcher (Chandrasekaran 1986), and failure handler. Figure (1) shows the relationship among these agents.

The Multiple Design approach (Kamel et al 1994) in general uses the same types of problem solving agents used by Routine Design. An additional type of agents called "design limiters" is also used in Multiple Design. However, the use of the problem solving agents in Multiple Design is different than their use in Routine Design.

For a given set of inputs (design requirements) a Multiple Design system generates a group of trees, where each node represents a value for a design attribute. Each path from a root to a leaf node represents a complete design. Figure (2) shows this form of output graphically (In this case a "design" is a set of values for attribute1, attribute2, and attribute3

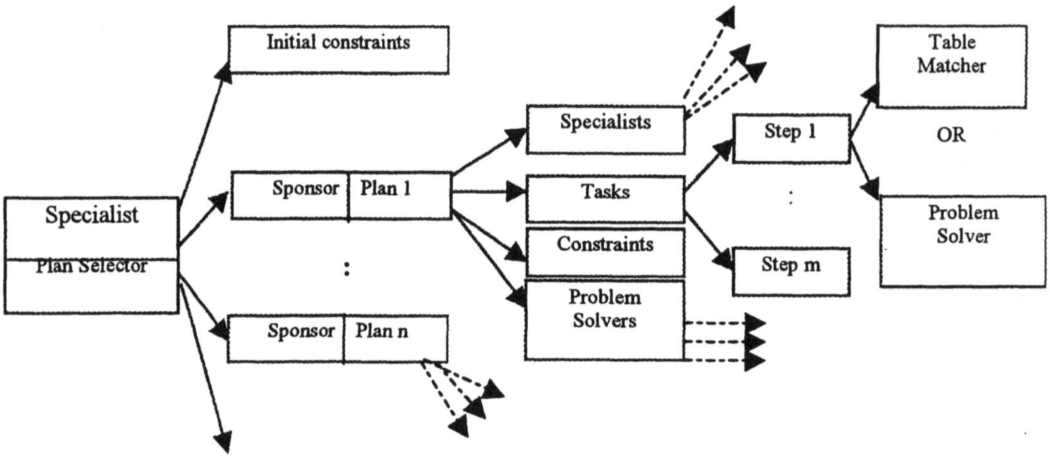

Figure 1 Relationship Among Routine Design Agents

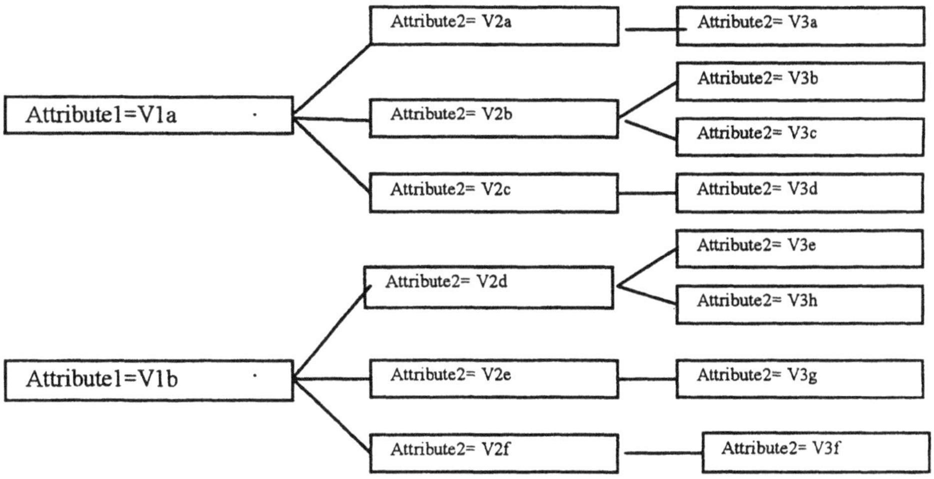

Figure 2 Graphical Representation of the Multiple Design outputs

3. CONCEPTUAL DESIGN OF NEPER EXPERT SYSTEM

3-1 NEPER Structure

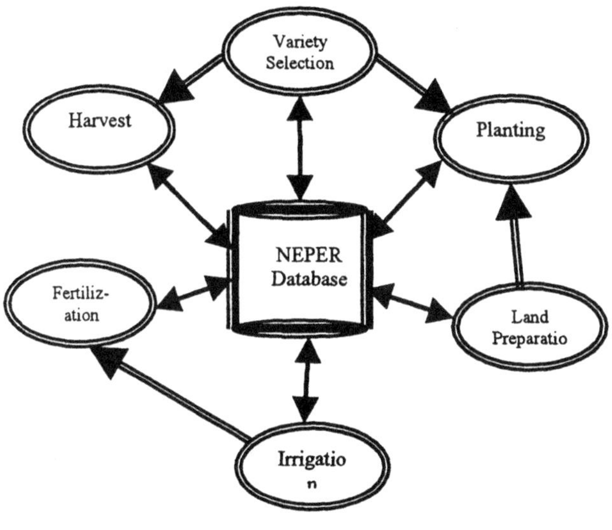

Figure 3 NEPER overall structure

NEPER Wheat Strategic Expert System consists of 6 subsystems namely: variety selection, land preparation, planting, irrigation, fertilization, and harvest. Each subsystem has its own database. There is also, common database of the whole system (NEPER), where the outputs of some sub-systems are used by others. Figure (3) shows the NEPER overall structure.

The common database includes both static and dynamic data. Static data contains all data that are rarely change (i.e. soil type, farm location) and those are not changed during season (i.e. soil and water salinity). Dynamic data contains all data that are usually changed during running NEPER.

The variety selection subsystem identifies the appropriate varieties for a specific site based on various parameters such as the soil type, soil salinity, drought, the weather, resistance to certain disease and others. The output of this subsystem is used by other subsystem such as planting and harvest subsystems.

Land preparation subsystem gives recommendation on how to prepare soil for wheat cultivation. This recommendation includes soil tillage, maintaining drainage system, getting red off previous crops and summer weeds, soil leveling, soil fine, and others. The output of this subsystem is used by planting subsystem.

Planting subsystem determines the appropriate planting date, planting methods, and seed rate. It uses the outputs of other subsystems (variety selection, land preparation) as inputs.

Irrigation subsystem gives a schedule plan for irrigation quantity, intervals, and irrigation time taken into consideration soil type, soil salinity, water quality, rain, temperature, for each specific site. Fertilization subsystem gives fertilization regime in terms of fertilizer name, dose, and application time, according to the soil fertility, previous crops, water quality, planting type, ..etc. Harvest subsystem gives a recommendation about the appropriate harvest date and used harvest machinery.

3-2 Knowledge Representation

NEPER Expert System has six-problem solvers to satisfy its plans namely: variety selection, planting, land preparation, irrigation, fertilization, and harvest problem solving methods. Figure (4) shows these problem solvers. Each problem solver has its own agent hierarchy. The top node of this hierarchy represents the top specialist, which in turn has its own plans. Each plan in this specialist has a set of tasks, each task has its own steps.

In our domain each subsystem has a set of strategic plans, each plan has a set of agricultural operation to be performed. The operation has a set of properties, which has single or multiple value (i.e. operation date, material, tool, method, and cost). In routine design the operation is represented as task and its properties are represented as steps. Assigning value for each property is controlled by a table matcher, and table (1) shows an example of land preparation knowledge, which assigns values for method property in tillage operation.

For example in Land Preparation specialist has one plan, which in turn has so many operation such as tillage operation, soil fine operation, maintaining drainage system, leveling, etc. Tillage operation for example, has four steps namely: tillage date, tillage equipment, tillage method and tillage cost. Figure (5) shows Land Preparation specialist.

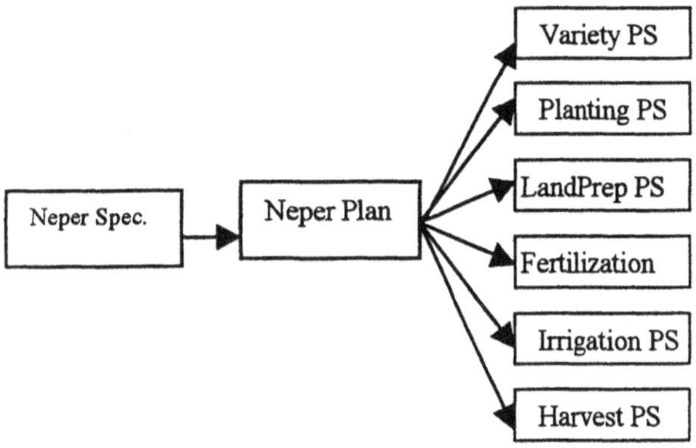

Figure 4 NEPER Problems Solving Methods

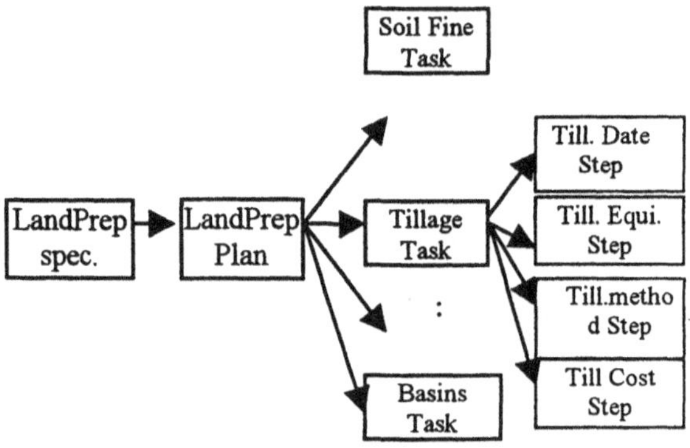

Figure 5 Land Preparation Specialist

Table (1) Example of land preparation knowledge

	Region	Soil Type	Plantin g Type	Chisel Availab le	Result
1	~=rain fed area	=sandy calcar	=Wet	~=No	Plow once and plowing depth is 15 cm
2	~=rain fed area	=black	=Wet	~=No	Plow twice and plowing depth is 15 cm
3	~=rain fed area	=calcareous	?	~=No	Plow twice perpendicularly and plowing depth is 15 cm

4 IMPLEMENTATION ASPECTS

NEPER Expert System was implemented using Generic Task tool (Sticklen et al 1992), which was built using Smalltalk language. This tool runs under VisualWorks (ParcPlace 1992) environment. It has a graphical user interface that facilitates the development of expert system. The tool contains three problem solvers namely: Routine Design, Hierarchical Classification, and Picture Classification. The Routine Design was used in developing the Strategic NEPER Expert System and it consists of three components namely database, agent hierarchy, and table matcher. These components enable the expert systems developers to develop their expert system.

The database component is used to define the required attributes, which are classified into three groups (input, intermediate, and, output groups) according to their use in the system. The input attributes are used to get the required values during running NEPER from the system users. The intermediate attributes are used to store the intermediate results during NEPER reasoning. The Output attributes are used to store the final conclusion of NEPER, which are displayed to the users. Four types of database were developed: farm data, soil and water data, equipment data, and fertilizer data. Figures 6, 6b, 6c, and 6d show snapshots from running the system for these databases.

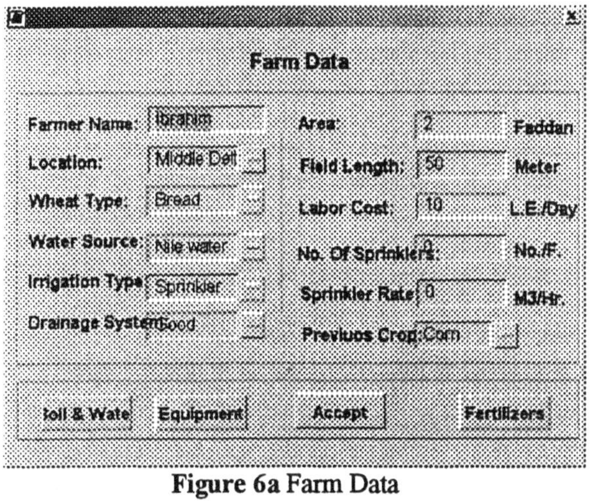

Figure 6a Farm Data

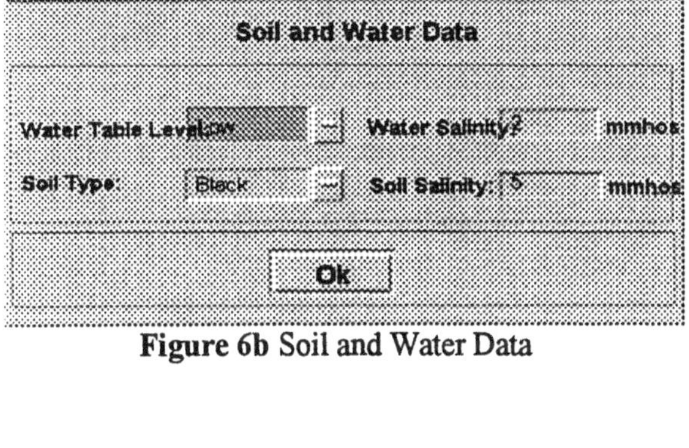

Figure 6b Soil and Water Data

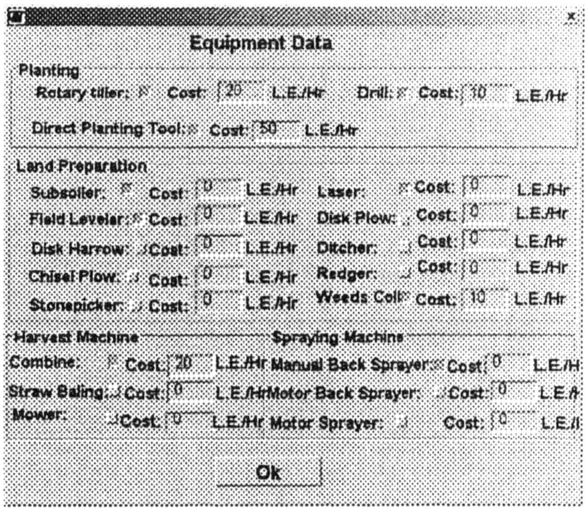

Figure 6c Equipment Data

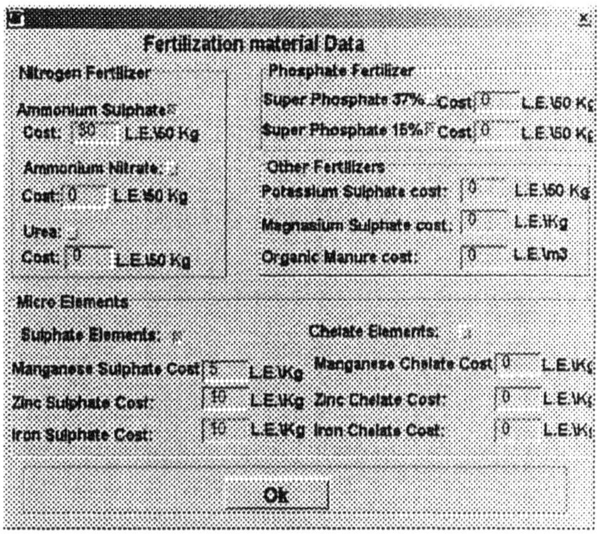

Figure 6d Fertilization Data

The agent hierarchy component is used to construct the NEPER agents' hierarchy. The NEPER has two types of agent hierarchy namely: NEPER hierarchy and problem solvers hierarchy. The NEPER hierarchy has one top specialist, which has main plan. This plan calls NEPER problem solvers hierarchy. These problem solvers are variety selection, land preparation, planting, irrigation, fertilization, and harvest. Each problem solver hierarchy is considered as a top specialist and main plan. This plan calls the problem solver tasks, which present the required operations. Each task includes a set of steps, which present the operation properties. These properties are assigned with their values according to the associated table matcher. The NEPER expert system works under Microsoft Windows and works in English/Arabic interface. The interface was developed using Smalltalk. The system runs under two types of running mode, single and multiple, modes. In single mode it gives unique solution, while in multiple mode, it gives more than one solution. Figure 7a and 7b show an example of the outputs screens for these types of running mode.

The system was validated by wheat scientists from International Center Agriculture Research in the Dray Area (ICARDA) and Wheat Division in Field Crops Research Institute (FCRI) and tested in six locations in Egypt. These locations cover most of the area that grow wheat and good results have been obtained against control field.

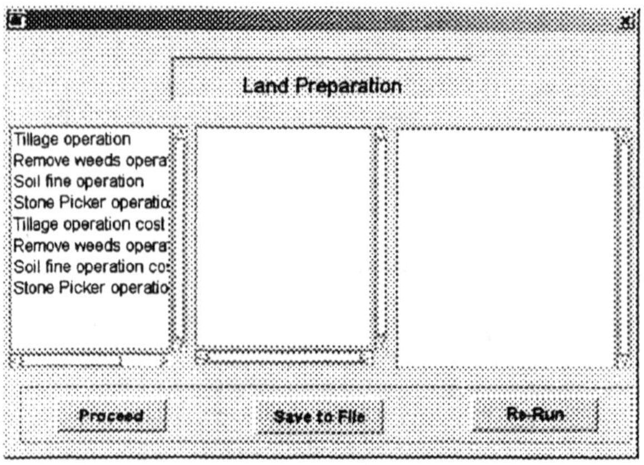

Figure 7a Single Output of Land Preparation System

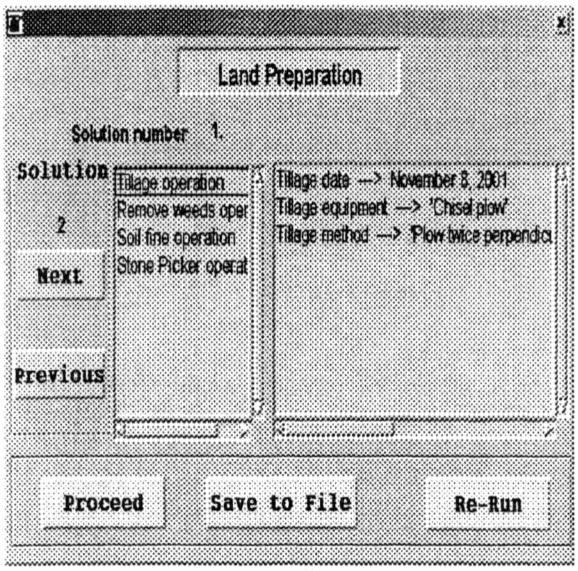

Figure 7b Multiple Output of Land Preparation System

5. CONCLUSION

NEPER wheat Expert System was developed, validated, and tested. It has been distributed for extension officers and growers. The system gives a strategic advice for the farmers, which is very important. This strategic advice avoid growers to have a problem during growing season, which in turn gives healthy plants with maximum product.

The developed system runs in English/Arabic language interface. Each sub-system runs alone in single/multiple mode. Running NEPPER in multiple mode calls all sub-system sequentially, which generates huge number of alternative solutions. These solutions can't be followed by the users and require huge memory to run. Therefore, it is recommended to run NEPER in single mode. If the user needs alternative strategic plans of a certain sub-system, he/she can run this sub-system in multiple mode.

NEPER is currently being enhanced to include an economic and environmental component to rank the alternative solutions. Integrating NEPER with CERES simulation model of Wheat growth and development (Ritchie, et al 1985) is also being considered.

REFERENCES

Chandrasekaran, B. (1986). Generic Task in Knowledge-Based Reasoning: High-Level Building Blocks for Expert System Design. IEEE Expert. 1: 23-30.

Kamel, A., McDowell, J., & Sticklen, J. (1994a). Multiple Design: An Extension of Routine Design for Generating Multiple Design Alternatives. Artificial Intelligence in Design '94, Lausanne, Switzerland.

ParcPlace (1992) "Visual Work", Release1.0, ParcPlace Systems,999 EastArques Avenue, Sunnyvale, California 94086-4593.

Rafea A. Ahmed (1998), Egyptian Research Program for Developing Expert System in Agriculture 7th International Conference Computer in Agriculture, Orlando, Florida, USA October 26-30th, 1998 Sponsored by ASAE

Ritchie, J. T., Godwin, D. C., Otter-Nacke, S.(19985) CERES Wheat , A Simulation Model of Wheat Growth and Development. Collage Station, Texas: Texas A&M University Press.

Sticklen, J., Kamel, A., Hawley, M., & Adegbite, V. (1992). Fabricating Composite Materials: A Comprehensive Problem Solving Architecture Based on a Generic Task Viewpoint. IEEE Expert, 7(2), 43-53

Tong, C. (1991) "Representation: structure, function & constraints; Routine design.", D. Sriram (Eds.), Academic Press Artificial Intelligence in Engineering Design: Volume I

THE FUNCTION OF THE GEOGRAPHIC INFORMATION SYSTEM (GIS) IN PRECISION FARMING

Pecze Zs. - Neményi M. - Mesterházi P.Á. - Stépán Zs.

University of West-Hungary,
Institute of Agricultural, Food and Environmental Engineering
H-9200 Mosonmagyaróvár, Vár 2, Hungary
Tel./Fax.: +36 96 215 911/209
e-mail: aenginst@mtk.nyme.hu

Abstract: Examinations in connection with site-specific farming have been carried out by our institute since 1998. Precision farming is a way of agricultural production which takes into account the infield variability. A technology where the application - seeding, nutrient replacement, spraying etc.- is taken place to act on the local circumstances of a given field. The GIS (Geographic Information System) created by computing background makes possible to generate complex view about our fields and to make valid agrotechnological decisions. Our goal was to compare two systems for marking out the further research tasks, because in several cases there are misunderstanding among the researcher, and the information provided by given companies are often also incorrect. *Copiright © 2001 IFAC*

Keywords: global positioning system, data handling system, data processing, data transmission, computer application

1. INTRODUCTION

Nowadays the environment protection is more and more in focus. This tendency can be observed in case of agriculture because of the direct or indirect human healthy food consumption is especially emphasised. The question of economy is also important.
In West-Europe and in the USA the precision plant production system, which takes into consideration the in-field variability is well known and getting more and more popular. An increasing attention for this technology is present in Hungary as well (Mesterházi et al., 2001).
Precision farming is a way of farming which takes into account the infield variability. A technology where the application - seeding, nutrient replacement, spraying etc.- is taken place to act on the local circumstances of a given field.
The GPS (Global Positioning System) makes possible to record the infield variability as geographically encoded data. It is possible to determine and record the correct position continuously. This technology considers the agricultural areas, fields more detailed than previously therefore a larger database is available for the user. For storing and handling these data the application of a GIS (Geographic Information System) is essential. The GIS created by computing background makes possible to generate complex view about our fields and to make valid agrotechnological decisions. Using this technology it is possible to provide the optimal or near optimal nutrient (Csizmazia, 1993) and chemical amount and

the proper cultivation for each part of the field (Jóri and Erbach, 1998). Consequently we can be able to save money and to prevent the environmental pollution caused by the leaching out of the nutrient and by the overuse of chemicals (Pecze et al., 2001). Examinations in connection with site-specific farming have been done by our institute for some years. To attain this technology there are several systems in the market, however at the present their reliability is poorly known by users and even by researchers in Hungary and abroad as well (Neményi et al., 2001).

2. MATERIALS AND METHODS

2.1 Description and circumstances of the experimental field

The experimental field is situated next to Ács (Komárom-Esztergom county) its topographical number is 0419/18-27, its area is 32.5 ha. The field shows variability in connection with relief as well, there is a 10 m of level difference.
The investigations were carried out in 1999 and 2000 after the pre-investigations in 1998. The examined plant was maize (Dekalb 443 and Dekalb 391 hybrids). The applied fertiliser is ammonium-nitrate (34%).
The harvest took part in 26-27 October 1999 and 9-13 October 2000, respectively.

2.2 Engineering background

The first task is make the machines able to record and transfer the proper information using the central control system i.e. to the automatic application of the decision based on that information. The other role is to equip the machines with fittings, which can modulate its operation parameters corresponding to the central control.

2.3 Informatics and GIS background

Our GIS's fundamentals are the ArcView based AGRO-Map Professional v.3.0 and the MapInfo v. 6.0 software. The board computer mounted on the harvester, on the soil sampler unit and on the applicator machine. The software installed on the computer are also the part of this system. The co-ordinates and the attributum data (yield-, soil physical and soil chemical properties, fertliser amounts etc.) are linked by the board computer. During the data gathering (yield mapping, soil sampling) the RDS and the Agrocom systems were used. For the visualization of the collected data the same systems' software were used designed for GIS applications. For data transfer floppy disk and PCMCIA card were used.

2.4 Information collected from primary and secondary data gathering methods

Two types of data collection were used. These are:

Data gathered by primary method:
- Field contour
- Yield map
- Soil supported maps
- Digital relief map
- 1:4000 black and white airborne picture
- 1: 10000 black and white airborne picture and the scanned, digitised form of it

Data gathered by secondary method:
- 1:10000 genetic soil map with the lab data and explanation text
- 1:10000 land use map, 1:4000 land use map
- 1:10000 scanned, digitised topographic map
- field register book

3. RESULTS

3.1 Yield maps

Data are recorded in regular ASCII format. After the data imported from the floppy disc (RDS system) into the personal computer the picture with the harvest route map were created as row data (Fig. 1).

Fig.1. Harvest route map for raw data (RDS, 1999)

Following this, using the geographical encoded data recorded in each 3 minute the interpolated yield map shows the infield variability (Fig. 2).

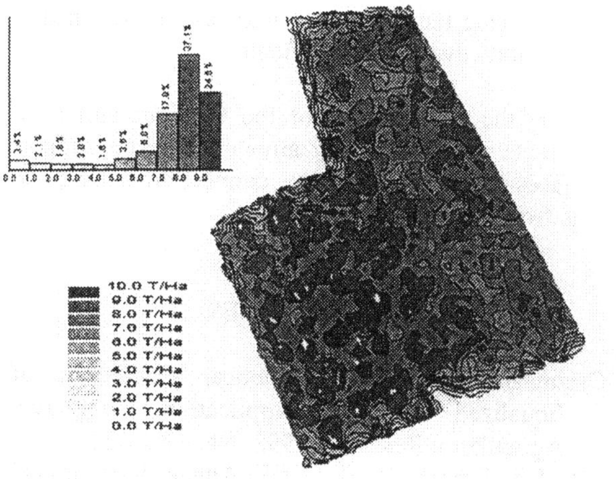

Fig.2. Yield map (RDS, 1999) The histogram shows the distribution of the yield categories

3.2 Soil maps and digital relief model

After grid soil sampling (1999) and the analysis the MapInfo Professional program was applied for the visualization. The point allocation (2000) were carried out by means of soil sampling plan generated by the AGRO-MAP Basic program. After the soil-analysis the soil maps were created from the soil nutrient-content properties as attributum data connected with the sample points (Kriging method). For building up of the relief model the co-ordinate data of the RDS PF data basis are originated from the harvest and the MapInfo Vertical Mapper software were used (Fig. 3).

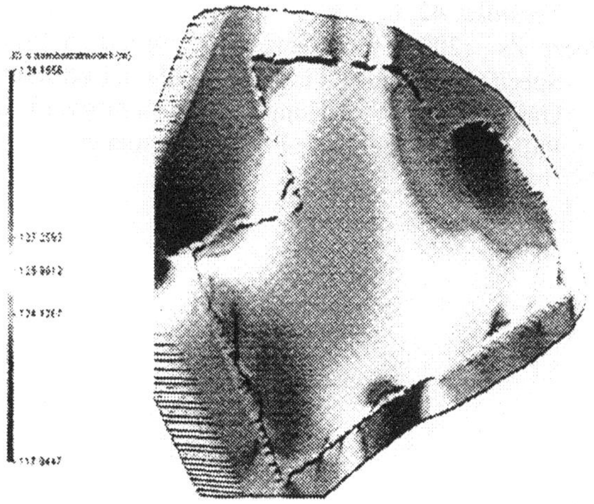

Fig. 3. The digital relief model of the field (MapInfo Vertical Mapper)

3.3 The data processing collected by secondary method

The 1:10.000 land use map in paper is prosured from the Region Land Office, Komárom. After scanning and knowing the co-ordinates of the field borderline, 4 points with known co-ordinates were identified in the map by means of a program called Wgeo. By doing this a .twf file - a file belongs to the .tif file - was generated. The AGRO-MAP Professional software has the transformation equation which is required for the transformation from WGS 84 GPS co-ordinates to EOV (Uniform Nationwide Projection System) thus the border line co-ordinates recorded by GPS could be handled in the EOV system. The database of the map with .tef extension were imported into the AGRO-MAP Professional program and the field borderline could be inserted in the map as a new layer (Fig. 4).

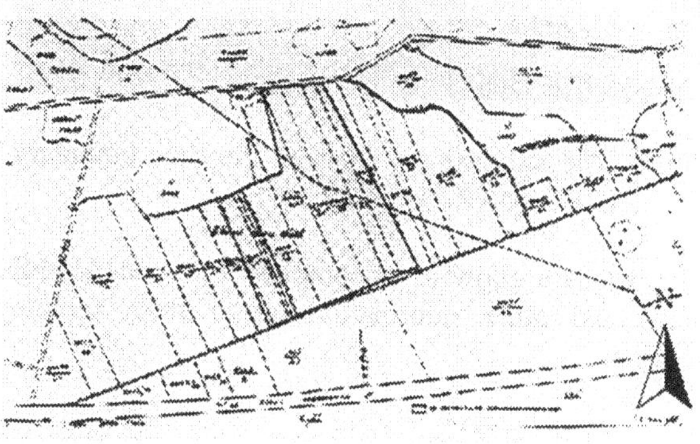

Fig.4. Land use map with the field border line (AGRO-MAP Professional) (The investigated area is dark-grey.)

The 1:10.000 topographic map is obtained from the Institute of Geodesy, Cartography and Remote Sensing, Budapest. This map imported into the AGRO-MAP Professional software composes an additional layer about the given field (Fig. 5).

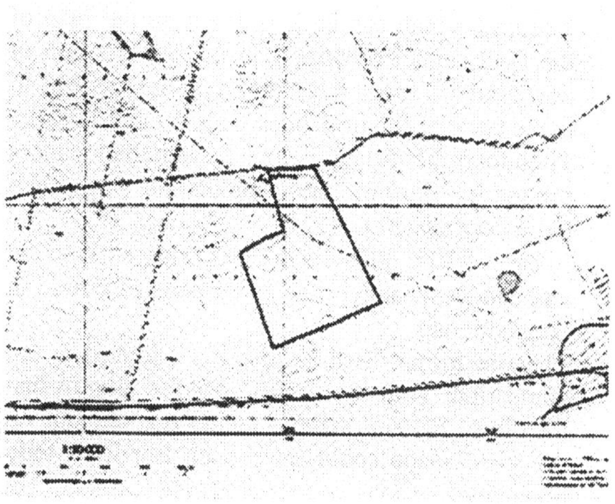

Fig. 5. The topographic map with the field border line (AGRO-MAP Professional)

The 1:10.000 scanned digitased black and white airborne picture is also originated from the Institute of Geodesy, Cartography and Remote Sensing, Budapest. By means of the Wgeo program a new layer was created from that in the same way like in case of the land use map (fig. 6.).

Fig. 6. The airborne picture with the field boundary (AGRO-MAP Professional)

The database allow us to complete with genetic soil maps and other information origin from remote sensing.

4. EXPERIENCES

Our experiences drows the attention to the fact that despite the standardization efforts the data from the GIS software for agricultural applications cannot be importet by simply conversion into the widely applied topographic programes (ArcView, ArcInfo, MapInfo etc.). Whereas, the following problems also exist in case of the GIS softwares for agricultural application:

- The grid which can be laid down on the map of the field cannot be turned during the creation of any kind of map, therefore it is impossible to make parallel the grid lines with the borderlines. Therefore the fitting is not perfect that causes problems during the application as well (Agrocom system).
- In case of RDS PF software it is impossible to cut and handle separately the given parts of a field in the yield map.
- The file format used by the RDS system is not compatible with the AGRO-MAP's file format therefore, the soil sample points marked out on the yield maps could have been imported only manually.
- The accurate yield date can be reported only in the points where GPS position recording has happend. In any other case information is provided in intervals with 1t/ha steps.
- During fertiliser application the ACT unit mounted in the cab of the tractor showed the position of the machine but did not provide directional guidance in spite of the fact that the working width was pre-programmed.
- Further problems were caused by that the ACT does not show the already distributed fertiliser strip in the field. This increases the chance of fertilising the same spot twice as well as making accurate overlapping difficult.

During the development of the GIS our aim is to procreate the compability among the export/import surfaces in other words the connection among the data from different sources.

5. REFERENCES

Csizmazia, Z. (1993) Technical Conditions of Equalized Fertilizer Application. Hungarian Agricultural Research, 1993, **No. 4** P16-22.

Jóri, J. I., Erbach, C., D. (1998) Annual International Meeting Sponsored by ASAE, **Paper No. 981051**

Mesterházi, P.Á., Pecze, Zs., Neményi, M. (2001). A precíziós növényvédelmi eljárások műszaki - térinformatikai feltételrendszere (The engineering and GIS background of the precision farming technology). *Növényvédelem*, in print

Neményi, M., Pecze, Zs., Mesterházi, P.Á., Kiss, E. (2001). Engineering environment of the precision crop production *Hungarian Agricultural Engineering*, in print

Pecze,Zs. - Neményi,M. - Kiss,E. - Petróczki,F.: 1999: Investigation of the accuracy of the RDS yield mapping system. 2nd European Conference on Precision Agriculture Abstracts. Odense, Denmark. 11-15. July.

Pecze, Zs., Neményi, M., Mesterházi, P.Á. (2001). A helyspecifikus tápanyagvisszapótlás műszaki háttere. (The technical background of the site-specific nutritient replacement) *Mezőgazdasági Technika,* **42**, 02:5-6.

Pecze Zs., (2001) Casemaps of the precision (Site-Specific) farming. Thesis of PhD Dissertation, University of West-Hungary, Mosonmagyaróvár, PhD Thesis supervisor: Prof. M. Neményi.

DISCRETE EVENT SIMULATION AS A PERFORMANCE ANALYSIS TOOL IN AGRICULTURAL LOGISTICS SYSTEMS

Daniel Nilsson

*Department of Agricultural Engineering,
Swedish University of Agricultural Sciences,
P.O. Box 7033, S-750 07 UPPSALA, Sweden*

Abstract: This paper presents a discrete event simulation model for analysing the performance and costs of straw and reed canary grass (RCG) logistics systems. The straw and RCG were used as fuels in a district heating plant for generation of hot water. The model, which was called SHAM (Straw Handling Model), follows the fuels from the fields via stores and heating plant to delivery of hot water. Some results from simulations with different machinery combinations, fuel proportions and storage capacities are presented, together with a discussion about advantages and limitations when applying this type of simulations on agricultural logistics systems. *Copyright © 2001 IFAC*

Keywords: discrete event dynamic systems, simulation, agriculture

1. INTRODUCTION

1.1 Background

Extensive allocation and concentration operations in space (transports) and time (storage) are necessary in agricultural production. Extensive transports are required because the farm products are produced over large areas and in many cases far away from the consumers. Most products also have low bulk densities and low dry matter contents. Furthermore, the point of time for harvest and use of farm products usually do not coincide, or are at least not carried out at the same rate, which necessitates storage facilities. Many, more or less, un-controllable factors must also be taken into consideration, for example, the weather and its impacts on soil properties, crop growth and quality of final product.

Harvest and handling of autumn-harvested cereal straw and spring-harvested reed canary grass (*Phalaris arundinacea L.*) for fuel purposes is an example of such logistics systems. These biofuels are produced over large areas, have low bulk densities, and must be harvested during a short period to avoid high moisture contents and fulfil the heating plants' fuel quality demands. Thus, a well thought-out handling system is necessary to provide the heating plants with these fuels at competitive costs. Co-handling of straw and reed canary grass (RCG) is in particular advantageous from an economic perspective, because the same machines, stores and other equipment can be used as these fuels are harvested at intervals of about six months. For a given heating plant, the lowest fuel costs will appear at an optimal number and combination of machines, at optimal location, number and space capacity of stores, and at optimal proportions of these fuels.

Discrete event simulation is an interesting approach to model such complicated logistics systems. The conceptualisation of the system into a model makes it possible to compare the performance and cost potential of different operating strategies before they, if promising, are realized into operable systems. Discrete event simulation is in particular useful to identify bottlenecks in resource-competing dynamic and stochastic systems. Typical applications of such simulations include transport systems, manufacturing plants, computer networks, service operations, etc., all including man-machinery interactions.

Several discrete event simulation models applied on agricultural systems were found in literature, especially for simulation of hay-making systems (Russel, *et al.*, 1983; Savoie, *et al.*, 1985; van Elderen, 1987; Buck, *et al.*, 1988; Axenbom, 1990; Gupta, *et al.*, 1990). These models could, however, not directly be adapted for simulation of straw and RCG logistics under Swedish conditions. New simulation languages with graphical user interface and powerful verification and animation tools have been developed in recent years. Therefore, a discrete event simulation model, based on the SIMAN/Arena language (Kelton, *et al.*, 1998), was developed in this project to simulate handling of straw and RCG.

1.2 Objectives

The objective of this study was to analyse the performance and costs of various logistics systems for handling of straw and RCG. These fuels were to be used in a district heating plant for generation of hot water. The tool for the analyses was a dynamic simulation model, called SHAM (Straw Handling Model), which was based on the discrete event simulation approach. Advantages and limitations with this type of simulations were also discussed in the study.

2. MODEL DESCRIPTION

2.1. Discrete event simulation and the SIMAN/Arena language

Discrete event simulation is defined as the modelling of a system as it evolves over time by a representation in which the state variables change instantaneously when events occur at separate points in time (Law and Kelton, 1991). Thus, the state variables are asynchronously driven by deterministic or stochastic events (figure 1).

The simulations in this project were implemented in the SIMAN/Arena language, which is based on the process-oriented simulation scheme. In such languages, entities, representing e.g. bales, vehicles, etc., flow through the system requesting service from resources. As the entities flow, they are delayed in queues and during processing. Each entity has its unique characteristics stored in an attribute array, which dynamically changes as the entities undergo the processes. Characteristics of the system as a whole are stored in global variable arrays (Kelton, *et al.*, 1998).

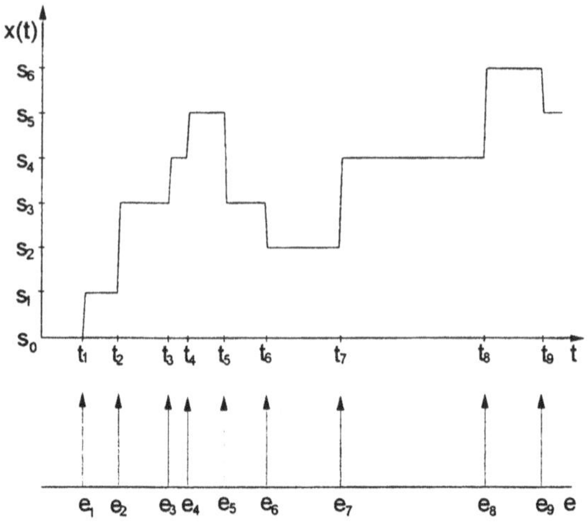

Fig. 1. An example of a trajectory in a discrete event system. The state *s* of the system is changed when event e_i occurs at time t_i.

The logic structure of SHAM was constructed by building networks of so-called modules through which the entities flow. The modules represent arrivals, departures, resources, counters, queues, assignments of attributes or variables, transporters, etc., and generate the underlying programme code when appropriate data are filled in. The modules belong to panels containing modules for different simulation applications and hierarchical levels.

2.2. Description of SHAM

SHAM simulates centrally-managed handling systems. It consists of a number of submodels (figure 2), which are run simultaneously, except for the location submodel, which is run before a simulation, and the cost/energy submodel, which is run after a simulation.

The location submodel uses geographical data as inputs to calculate transport distances between fields and stores or other fields, and to calculate field areas.

The straw moisture content is predicted in the drying submodel. It uses historical weather data to calculate the moisture content at any time by means of a semi-empirical continuous simulation model. The moisture content should not exceed 18% (w.b.) in order to avoid stoppages in baling machines and feeding equipment at the heating plant, and to avoid fungal growth during storage.

The soil status submodel uses historical weather data and data on soil properties to predict the soil moisture contents. This submodel is necessary in order to avoid significant soil compaction by the machines.

The growth submodel calculates the height of the new RCG shoots in the early spring by means of a regression equation, which uses a temperature degree sum as input. RCG should not be harvested too late, because cutting of new green tops will increase contents of moisture and minerals in the fuel, and may also affect yield in subsequent years.

The daily fuel use submodel is used to determine inventory levels and storage capacity necessary. It computes the daily fuel use in the heating plant by using historical weather data and technical information about the heat conversion system, as well as the distribution and consumer systems.

The submodels described above transfer their information to the handling submodel, which is the basic part of SHAM. Various machinery combinations and management strategies can here be simulated to evaluate their performance in handling of straw and RCG. The performance results are used in the final submodel to calculate the costs, labour demand and energy requirements for the simulated scenarios. This module is written in a spread-sheet programme.

SHAM and its submodels are described in greater detail by Nilsson (1999a).

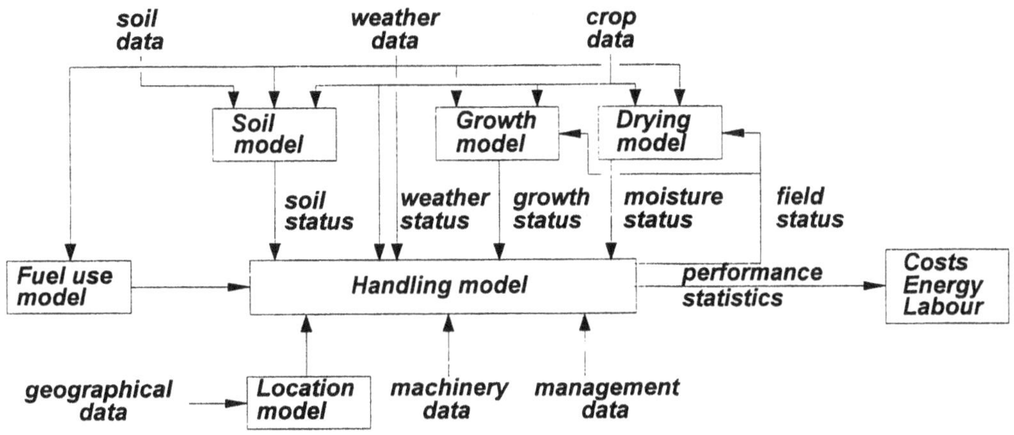

Fig. 2. Overview of SHAM and its submodels (Nilsson and Hansson, 2001).

2.3. Performance measures and costs

Several measures of system performance can be recorded in SHAM, depending on the purpose of the study. Some examples are: the annual quantity of fuels delivered to the heating plant, which may be seen as a consumer-oriented measure, or the utilisation and idle times of machines, which may be seen as a contractor-oriented measure, or the time the fields are occupied with un-harvested straw, which may be seen as a farmer-oriented performance measure.

The costs are in most studies the most important output measure. Calculation procedures and assumptions are described by Nilsson (1999a).

2.4. Verification and validation

SHAM was verified by walkthroughs, animated test runs and entity flow tracing. The validity of the model as a whole was established by face validation with persons familiar with the system. The validity of the submodels describing field drying, early growth of RCG, and soil status, was tested by comparisons with results from field experiments (Nilsson, 1999b; Nilsson and Hansson, 2001).

3. SIMULATION ASSUMPTIONS

The study was carried out for a fictitious district heating plant in Enköping (59°40N, 17°00E) in central Sweden. It was assumed that the heat consumers had a demand of 12.0 GWh for space heating and 4.0 GWh for hot water preparation in a normal year, and that the distribution losses were 20%. The base load boiler, which was solely fuelled with straw and RCG, had a maximum power output of 4.0 MW. Oil was used as a peak and reserve load fuel.

The fuels were harvested by a number of high-density balers, which produced bales weighing

540 kg. The bales were collected and loaded on transporters by a loader in the field, and unloaded and stacked by a loader at the storage. It took 30 minutes for them to load or unload the transporters, which carried 24 bales and had an average transport speed of 20 km/h. The location of fields and the three stores were determined according to prevailing geographical conditions. The machine capacities were obtained from time studies described in the literature.

Winter wheat was used as the straw-producing crop. The yields per hectare of straw and RCG were determined from probability distributions, as well as the start of the combining season and the point of time when a certain field with winter wheat was combined, and also the time between and duration of machinery stoppages,

A sandy clay loam and a clay loam were assumed to be typical representatives of the soils in the fields where RCG and winter wheat were grown, respectively. A field was trafficable when the soil moisture contents were within specified limits, and when there was no snow (for harvest of RCG). The baling operation was not allowed if the moisture contents of the fuels were higher than 18% (w.b.) or when the fields were not trafficable. It was assumed that harvest of RCG should not be performed later than the point of time when the new shoots were higher than 12 cm. In this study, however, all fields with RCG were harvested regardless of the height of the new shoots. The quantity of RCG harvested "too late" is instead presented in the results.

The primary fuel costs for oil, RCG and straw were: 90.6 SEK per GJ delivered heat, 21.0 SEK per GJ delivered heat (including cultivation and mowing costs), and 5.9 SEK per GJ delivered heat (paid to the farmer for the un-baled straw located in the field), respectively.

Fifteen replications, i.e. the years 1980-94, were simulated for each scenario, using the common random number technique with complete synchronization in order to reduce the variance.

4. RESULTS

The simulations showed that the total heat demand ranged from about 65 GJ/year in 1990 to about 79 GJ/year in 1985. The daily heat loads required for two "normal" years, 1982 and 1983, are shown in figure 3. The weekly inventory levels for a system with three balers, four transporters, three stores having a storage capacity of 2 000 tonnes each, and a cultivation area of RCG on 285 ha, corresponding to an average harvest of RCG of 2 000 tonnes/year, are also shown in the figure.

During 1982-83, the stores were (almost) empty only one week in March 1982, before the machines started to harvest RCG (figure 3). Furthermore, the stores were full only one week in September 1983, when the machines had to stop their harvest of straw. It can also be seen that the capacity of the machinery chain was too low in the autumn of 1982, because the stores were not fully replenished with sufficiently dry straw this year.

The costs for systems with two, three and four balers, as function of average harvest of RCG and storage capacities, are shown in figure 4. The average quantity of RCG harvested when the new shoots were higher than 12 cm during the fifteen years simulated is also shown. Four transporters were used in all these simulations. The reason was that the transporting team worked independently of the balers, and it was found that the field loader and the store loader, regardless of the number of balers used, should work with at least four transporters in order to reduce loader waiting times.

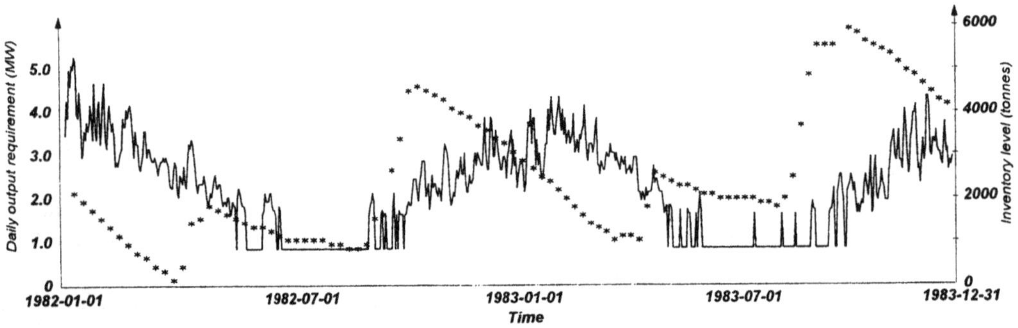

Fig. 3. Daily heat requirements (solid line) during the years 1982-83. The weekly inventory levels (*) for a system with three balers, four transporters, a storage capacity of 6 000 tonnes and an average harvest of RCG of 2 000 tonnes/year are also shown (Nilsson and Hansson, 2001).

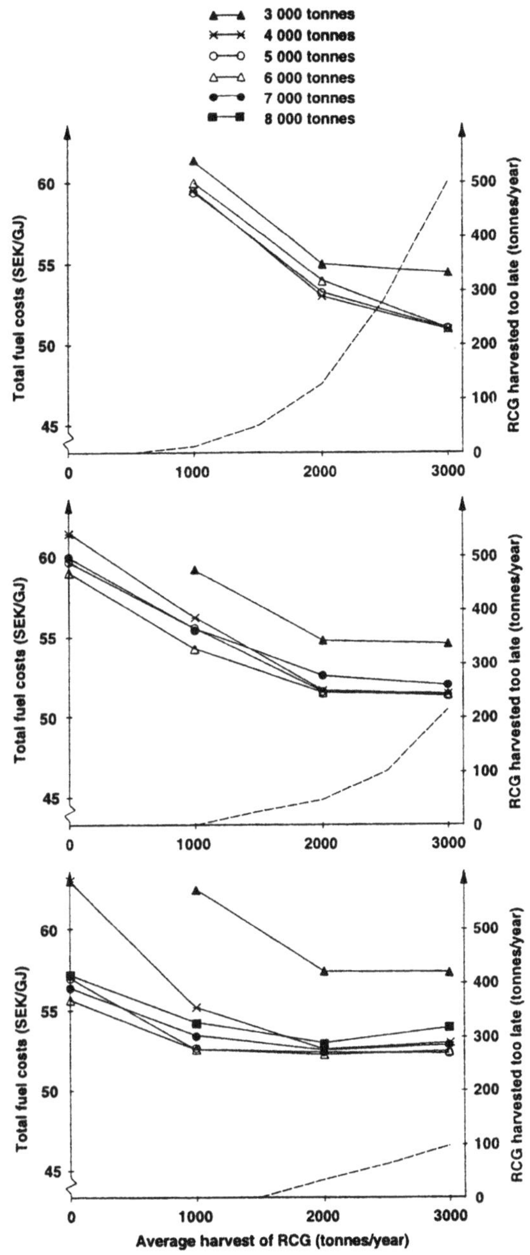

Fig.4. Total fuel costs in SEK per GJ delivered heat for systems with various storage capacities (solid lines) and two (top), three (middle) and four balers (bottom), and average quantity of grass in tonnes/year harvested when the new shoots were longer than 12 cm (dashed lines), as functions of average grass harvest (Nilsson and Hansson, 2001).

If no RCG at all was accepted to be harvested too late, then the lowest cost, 52.4 SEK per GJ delivered heat, was for a system with four balers with a storage capacity of 6 000 tonnes and an average harvest of 1 500 tonnes of RCG, corresponding to an RCG cultivation area of about 210 ha. During the fifteen years simulated with this system, about 10% of the heat delivered from the heating plant arose from the peak and reserve load fuel (oil), about 24% arose from RCG and about 66% from straw.

If a maximum of 100 tonnes of RCG per year was accepted to be harvested late, i.e. when the new shoots were higher than 12 cm, the lowest costs now appeared for a system with three balers, 5 000 tonnes of storage space and, on average, 2 500 tonnes of RCG harvested per year on 360 ha. This cost was 51.4 SEK/GJ, and about 9% of the heat delivered from the heating plant arose from oil, about 40% from RCG and about 51% from straw.

5. DISCUSSION

The costs for large-scale harvest of straw in the municipality of Enköping are about 20% higher than in southern Sweden, due to longer transport distances, lower and more variable yields of straw, and poorer weather conditions during the harvest season (Nilsson, 1999c). The simulations have shown, however, that the costs can be reduced by using RCG as a complementary fuel in straw-fired district heating plants, provided that appropriate machinery combinations, fuel proportions and storage capacities are used. The primary fuel cost for RCG, i.e. the cultivation and mowing cost, is more than three times higher than for straw, but this is, to a certain degree, compensated by more effective use of machines and storage buildings.

The lowest cost at all was found when two balers were harvesting 3 000 tonnes of RCG per year. The quantity of RCG harvested too late was, however, unacceptably high (17%). There was no penalty cost in this study for harvesting RCG too late, because of the difficulties to estimate such a value.

The simulations showed that an increased number of balers is beneficial when the proportions of RCG are small, because the disadvantage of additional machinery costs are outweighed by the fact that it is possible to harvest more straw. It was also shown that suitable storage capacities for systems with three and four balers were as high as 6 000 tonnes. This value reflects that it is more profitable to have high buffer capacities instead of using large quantities of oil in years with poor harvest conditions.

The discrete event simulation approach proved to be a powerful tool to evaluate the performance of various alternatives in order to design cost-efficient system configurations. The discrete event simulation approach makes it possible to consider dynamic and stochastic system properties. Thus, it is possible to identify bottlenecks and critical points in handling systems like these.

The modelling in this project was also facilitated by the graphical user interface and the powerful model verification tools and animation capabilities in SIMAN/Arena. A problem was, however, to model the flexible and adaptable decision-making of humans. The modelling of such issues is complicated in process-oriented languages like SIMAN/Arena. Other difficulties were to establish feasibility limits for e.g. field operations, and that it was necessary to include continuous simulation to model the field drying of straw.

REFERENCES

Axenbom, Å. (1990). A simulation model for planning of hay harvesting machinery systems and management. Department of Agricultural Engineering, Swedish University of Agricultural Sciences, Uppsala.

Buck, N. L., Vaughan, D. H. and H. A. Hughes (1988). A general-purpose simulation program for agricultural operations. *Comp. and Electr. in Agric.*, **3**, 29-44.

Gupta, M. L., McMahon, T. A., McMillan, R. H. and D. W. Bennet (1990). Simulation of hay-making systems: Part 1 – Development of the model. Part 2 – Application of the model. *Agric. Systems*, **34**, 277-318.

Kelton, W. D., Sadowski, R. P. and D. A. Sadowski (1998). *Simulation with Arena.* WCB McGraw-Hill, New York.

Law, A. M. and W. D. Kelton (1991). *Simulation, modelling and analysis.* 2nd edition. McGraw-Hill, London.

Nilsson, D. (1999a). Analysis and simulation of systems for delivery of fuel straw to district heating plants. Dissertation. Agraria 205. Swedish University of Agricultural Sciences, Uppsala.

Nilsson, D. (1999b). SHAM – A simulation model for designing straw fuel delivery systems. Part I. Model description. *Biomass and Bioenergy*, **16**, 25-38.

Nilsson, D. (1999c). SHAM – A simulation model for designing straw fuel delivery systems. Part II. Model applications.. *Biomass and Bioenergy*, **16**, 39-50.

Nilsson, D. and P-A. Hansson (2001). Influence of various machinery combinations, fuel proportions and storage capacities on costs for co-handling of straw and reed canary grass to district heating plants. *Biomass and Bioenergy*. (In press).

Russel, N. P., Milligan, R. A. and E. L LaDue (1983). A stochastic simulation model for evaluating forage machinery performance. *Agric. Systems*, **10**, 39-63.

Savoie, P., Parsch, L. D. Rotz, C. A., Brook, R. C. and J. R. Black (1985). Simulation of forage harvest and conservation on dairy farms. *Agric. Systems*, **17**, 117-131.

van Elderen, E. (1987). *Scheduling farm operations: a simulation model.* Pudoc, Wageningen.

THE EVALUATION AND IMPACT OF NEPER WHEAT EXPERT SYSTEM

Ahmed Rafea

Computer Science Dept., AUC
Email: rafea@aucegypt.edu

Mostafa Mahmoud

Central Lab. for Agriculture Expert System, ARC
Email: mostafa@ esic.claes.sci.eg

Abstract: This paper presents the laboratory and field evalation results of NEPER Wheat expert system. The laboratory evaluation showed that NEPER performance is comparable with human experts. Field evaluation has revealed that NEPER has good economic and environmental impacts. The field testing results have also shown that NEPER is usable, applicable and needed. *Copyright ©2001 IFAC*

Keywords: Expert Systems, Diagnosis, Knowledge-based systems, Hierarchical structures, Classification, Intelligence.

1. INTRODUCTION

Bread, known as aish, or life, is a vital component of the Egyptian diet. In 1993, the country produced 4.5 million tons of wheat on 2.2 million feddans. Given the crucial role wheat plays in Egypt, CLAES cooperated with the Intelligent Systems laboratory (ISL) at Michigan State University in developing the Egyptian Regional Wheat Management System, funded by NARP a United States Agency for International Development (USAID) project in the period from 1992 to 1995. This project integrates an ES with a crop simulation model and aims at addressing all aspects of irrigated wheat management in Egypt. This integrated system is named NEPER (Kamel et al, 1995). In order to achieve this goal, NEPER is designed to perform the following functions:

- Select the appropriate variety for a specific field
- Advise the farmer on field preparation
- Design schedules for irrigation and fertilization
- Control pests and Weeds
- Manage harvests
- Diagnose malnutrition
- Diagnose disorders
- Suggest Treatments

In 1997, another project between CLAES and ISL was funded by the ATUT which is also a USAID project. One of the objectives of this project was conducting field-testing to measure the ES performance. The objective of conducting the field-testing was to evaluate the economical and environmental impacts and to measure the ES performance from three aspects: usability, applicability and need. The results of this testing were also used to enhance the user interface and extend the knowledge base of Neper .In this project, there is a component for the evaluation of a new enhanced version of NEPER that considers the whole agricultural operations. This new version has been developed according to the results and recommendation of the field testing that is presented in this paper.

In this paper, the technical background of NEPER is presented in section 2. The laboratory evaluation is summarized in section3. The experiments description is presented in sections 4. The economical and environmental impact of the ES are summarized in sections 5 & 6, respectively. The ES performance, section 7, has been measured using three aspects namely usability, applicability, and need of ES. ES enhancements as a result of those experiments are presented in section 8.

2. BACKGROUND

In developing Neper, the Generic Task Approach to ES development proposed by Chandrasekran (Chandrasekran, 1986) has been used. The idea behind the Generic Task approach, is that the way a problem is to be solved, depends largely on its type e.g., diagnosis, design, planning, etc. Consequently, problems of the same type could share some sort of a generic problem solver. So, according to the Generic Task methodology approaching a diagnosis problem will be inherently the same regardless of the domain in which such a problem is being addressed. The

classical example of a problem solver that could be applied to a diagnosis problem is Hierarchical Classification (Gomez & Chandrasekran, 1981; Chandrasekran, 1986) and it is this problem solver that has been used in implementing the Wheat disorders ES, which is a component of NEPER.
This system component has been implemented using a Generic Task Tool developed at Michigan State University (MSU). In this tool, the knowledge base is created as a hierarchy of nodes. In each node, the knowledge is represented in a table, where each entry in this table represents either a database variable or a variable pointing to another table. Each database variable is associated with a question. A user will be presented with that question only if the database variable has never been assigned a value. The combination of possible inputs for each question denotes different rules and matching patterns. If a combination of inputs results in a match value greater than a given threshold, the node is said to be established. By asking the user a series of questions, the system is able to pursue or rule out paths in the classification in which the leaves represent disorders. Basically, if a path from a root to a leaf exists, then the disorder at the leaf is presented as the diagnosis.

3. LABORATORY EVALUATION

Laboratory evaluation is conducted before dissemianting the ESs in the field. The Laboratory evaluation methodology consists of three main procedures namely Verification, Validation, and Evaluation. Verification is defined as the demonstration of consistency, completeness, and correctness of software (Adrion et al, 1982). O'Keefe et al. (1987, 1989, and 1990) have defined verification as "Building the system right", that is making sure that the implemented system is functionally matching the proposed design, and free of semantic and syntactic errors. Validation is the process whereby the system is tested to show that its performance matches the original requirements of the proposed system. It is defined as the determination of the correctness of the final program or software produced from a development project with respect to the user needs and requirements (Adrion et al, 19982). As noted by O'Keefe et al. (1987, 1989, and 1990) "Validation means building the right system". Evaluation is the process whereby we ensure the usability, quality, and utility of the ES (O'Keefe et al. 1987, 1989, and 1990). A complete testing cycle is performed in iterations through which, the ES is updated and refined.

Verification process evolves through two main stages during the development of the ES: the development stage and the examination stage. In the development stage, the developer practices different functions of the implemented systems, looking for potential errors that may exist. This is accomplished using two broad techniques: non case-based and case-based. Non case-based techniques include tracing, spying and other traditional debugging techniques. Case-based verification techniques are applied by preparing "Typical Cases". These cases should be selected to serve requirements satisfaction as spelled out in the requirement specification. In the examination stage, the ES is tested to make sure that it is running properly, by testing all the functions of the system trying to examine the performance of the system in different situations. The output of this stage is the *verification report* that is a document of differences between system design and implementation. This report is used to update , the design document and implementation.

The validation step is done through conducting meeting with the doamin experts who provided the knowledge to check that the right system has been developed. This is done by going throgh the generated test cases during the meeting with the domain experts. Their comments on the content and user interface are considered. Necessary updating of the design and implementation is done.

The evaluation step is to assess the quality, usability, and utility of the ES from the point of view of human experts other than the domain expert, who participate in the system development. Typical cases are created and distributed to three domain experts in the specialty of a specific sub system. If one sub system includes more than one specialty, cases are distributed to all experts in different specialties. For example in the remediation subsystem, we have three specialties: plant pathology, entomology, and nutrition. Therefore 9 experts have participated in the validation of this subsystem. For each specialty, an evaluator is selected to blindly assess the responses of the three human experts and the ES. After the evaluation, the domain expert participated in the development, the evaluator, and the domain experts participate in an evaluation meeting together with the knowledge engineer to discuss the evaluation results till they reach to a consensus.

Applying this methodology on NEPER, verification and validation were done sucessfully. In this paragraph we will presnt the evaluation results of the diagnosis and treatment subsystems. Figure 1 shows the evalaution scores of NEPER diagnosis subsystem. NEPER diagnosis over performs human expert in the insect and malnutrition speciltles, and its score in the disease diagnosis results (86%) is equivalent to those of the best human expert.

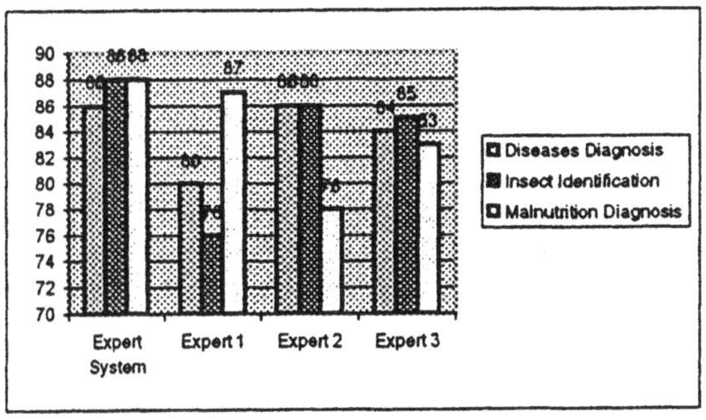

Figure 1 NEPER Diagnosis evaluation result

The evaluation scores of NEPER treatment subsystem is shown in figure 2. NEPER treatment over performs human expert in disease treatment, and its score in the insecta and malnutrition treatment are 0.95 and 0.85 respectivly of the best expert-group. After this experiment, NEPER has been trained to reach the scores of the best experts..

4. FIELD EXPERIMENT DESCRIPTION

Many experiments were conducted in the last few years for NEPER ESs. The objectives of those experiments were to validate the system in the field, and to measure the impact of using the system. The experiments were conducted in different locations by selecting two fields at same area and location: one is to be cultivated using NEPER ESs recommendations without any interference from the agriculture engineer or any specialist, and the other one is to be cultivated as usual, this is a control field.

In order to get the best results from the experiment, the following issues and activities were considered and followed:

♦ Formal training on the usage of NEPER was conducted for the staff who are going to use the system.

♦ A computer engineers from CLAES were responsible for supporting the site staff on the usage of the system and handling trouble-shooting problems of hardware and software.

♦ A number of the wheat researchers from Field Crop Research Institute (FCRI) were assigned to supervise different fields, i.e., a researcher for each site.

♦ Periodical fields visits were conducted by researchers from CLAES and FCRI

Three experiments were conducted in three different seasons for NEPER ES. Two of those three experiments had the same number of fields, both of them consisted of a total number of 32 fields carefully selected for conducting the experiment. The third one consisted of a total number of 44. These fields were equally divided so that 16 fields in the first two expermints and 22 fields in the third one were assigned to utilize NEPER and managed by the ES, and the other fields were to be managed in the usual practice and acts as control. The selected fields were located in four different geographical areas, namely: Noubaria, Gemiza, Sharkia, and Decerns. In Noubaria two sites were selected to cover the different types of soil at that area. One of those sites located in Bostan and the other one located in Banger El-Sokar.

The first experiment covered only the diagnosis and treatment part of the NEPER Wheat ES including Weed Identification. The second one included the strategic part and tactic part. Strategic part includes six subsystems called: Variety Selection, Pre-cultivation Pest Control, Tillage, Planting, Irrigation & fertilization, and Harvest. The third one also included the strategic part and tactic part. Strategic part includes six subsystems called: Variety Selection, Planting, Land Preparation, Irrigation, fertilization, and Harvest. Tactical part includes two subsystems called diagnosis and weed identification, each of them includes the treatment function.

5. ECONOMICAL IMPACT

In the first experiment (CLAES, 1996), the averages of treatment costs, yields, and straw per feddan was calculated for both NEPER and the control fields. By taking the averages of treatment cost, yield, and straw per Feddan, it was found that the average net income per feddan for ES fields is 2049.85 LE and for control fields is 1600.05 LE, consequently, the net production increase in Egyptian Pound was 449.8. This represents 26.78% increase in the production. In the second and third experiments (CLAES, 1999, CLAES, 2001), the complate system was tested. Tables (1) and (2) summarize the result of those experiments in the new reclaimed area and the Delta area. The following remarks were observed:

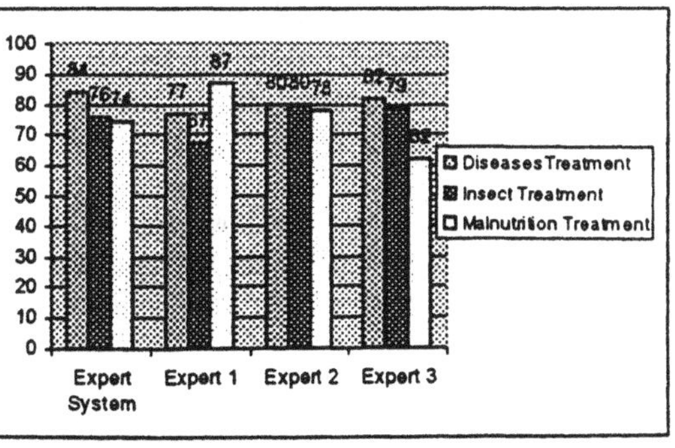

Figure 2 NEPER Treatment evaluation result

Table 1: The result of the experiments in the new reclaimed area

Item	Season 97/98				Season 98/99			
	ES	Control	Differance	%	ES	Control	Differance	%
Average Production	1701	1506	195	13	1647	1431	216	15
Average cost	747	861	-114	-13	468	603	-135	-22
Average Net profite	954	645	309	48	1179	828	351	42

Table 2: The result of the experiments in the Delta area

Item	Season 97/98				Season 98/99			
	ES	Control	Differance	%	ES	Control	Differance	%
Average Production	2130	1830	300	16	2117	1759	358	20
Average cost	631	597	34	5	445	422	23	5
Average Net profite	1499	1233	265	22	1672	1337	335	25

- In both the new reclaimed and Delta area, there was an increase in the production and net profit consistently in the two consecutive seasons.
- The percentage of increase in the net profit in the newly reclaimed is greater than the percentage of increase in the net profit in the Delta area.
- The production in the newly reclaimed area is less than the Delta area because the lack of expertise in the reclaimed area. Hence expertise transfer in this area has led to a relatively high impact.

6. ENVIRONMENTAL IMPACT

The conservation of natural resources has two aspects. The first is pertinent to the management of these resources on the macro level, such as controlling the expansion of urban development in order not to loose agricultural land. The second is concerned with the management of these resources on the micro level such as adding chemical fertilizers to the soil. In this paper, the focus will be on the status of the water and land resources because they are the two main resources related to our work on crop management ESs. Water is the scarcest resource in Egypt, since its supply is nearly fixed and water demand for different sectors is continuously increasing. The decision makers concerned with water resource management in Egypt are challenged by how to balance the limited water supply with an increasing water demand for the future, since water is the major constraint for land expansion to satisfy food self-sufficiency. Another challenge is how to reduce the water pollution resulting from using chemical fertilizers and pesticides. After water, land is the major limiting factor for sustainable agricultural development (Rafea, 1996).

There are two problems facing decision-makers to conserve water resources namely: the efficient utilization of water resources, and the pollution resulting from the usage of chemical fertilizers and pesticides. Regarding soil conservation, there are two main problems namely: the urban expansion, and the soil degradation resulting from excessive use of fertilizers and other bad agricultural practices. Therefore, the main contribution of ESs for soil and water conservation is to transfer the agricultural practices according to certain strategy or a combination of strategies namely: environmental sustainability, economical sustainability and/or social sustainability. In the ESs that have been built so far, we are concerned with economic sustainability taking into consideration the environmental sustainability in the second place. In other words, we are trying to acquire the recommendations that optimize the output relative to the agricultural inputs. As a consequence, environmental conservation is achieved, because no extra input is provided such as water, fertilizers and pesticides without a return in the yield.

The results of experiments conducted for the ES agree with the goals of environmental conservation. The fields managed by the ES have used fewer resources in terms of fertilizers and pesticides than the control fields and hence conserve environment. The cost is an indicator of the increase or decrease of using chemicals in general. Hence, we have used the cost as a factor in determining the quantity of used fertilizers and pesticides. The average cost of pesticides used by NEPER Wheat fields in the first experiment was more than the control fields by 15.7 Egyptian pound/Fadden, but the production increased by 449.8 Egyptian pound/Fadden. Notice that the increase of cost in this experiment is negligible. In the second experiment the average cost of fertilizers and pesticides used by NEPER Wheat fields was less than the control fields by 5.57 Egyptian pound/Fadden and the production increased by 247.4 Egyptian pound/Fadden. In the third experiment the average cost of fertilizers and pesticides used by NEPER Wheat fields was less than the control fields by 1.2 Egyptian pound/Fadden and the production increased by 287.12 Egyptian pound/Fadden. In fact, this indicates that changes to the NEPER Wheat system have made it more compliant to the goals of resource management and environmental conservation. In the second experiment of NEPER wheat ES, the average water quantity used to produce one Ardab of wheat in the ES fields was 112.74 M3 water, while in the control fields the farmers used 152.58 M3 water on average to produce the same quantity of wheat. This represents 35% decrease in the use of water.

7. EXPERT SYSTEM PERFORMANCE

The expert system performance has been measured using three aspects namely usability, applicability, and need of ES.

7.1 Expert Usability

In order to measure the usability of the ES, the developers in CLAES have re-run the system on the cases reported in the forms of the fields managed by NEPER and compared the conclusions with the results represented in the field books by the researchers and extension agriculture engineers in different locations. In the first experiment (CLAES, 1996) was examining the comparison results, it was found that in 86% of the cases, the trained researchers have used the system correctly while this percentage has decreased to 38% for untrained researchers. This indicates the importance of training on the usage of the ES. It is worth noting that there is no great difference between the researchers and extension officers in using the system as the differences was only 4%, although the system was in English. This proves the importance of ES. It raised the performance of extension officers to the level of researchers, in the underlying domain of the NEPER.

In the second experiment there was discrepancy between the ES recommendation and the agriculture practices documented in the field books. When this discrepancy was discussed with the ES users, we found that this discrepancy was due to their rejection of the ES recommendation and not due to bad use of the system. Therefor, we concluded that the usability of the system in the second experiment was high.

7.2 Expert system applicability

The applicability can be measured by comparing the ES recommendation and to what extent the ES users have applied them. This discrepancy must not be due to bad usability of the system.
In the first experiment, the discrepancy between the ES results and the applied practices by the users were due to bad use of the system.
In the second experiment (CLAES, 1999), it was difficult to quantify the comparison result as it was found that sometimes the recommendations are applied partially. Hence qualitative measures were found more appropriate, especially in the strategic part. The applicability of the modules: Pre-Cultivation Pest Control, Planting, and Weed Control was found low because in Noubaria fields' users did not accept the ES recommendations of the pre-cultivation pest control and the planting modules. In the weed control module the actual practice is different from the ES advice. The applicability of the modules: Tillage and Fertilization are moderate, as the ES fields' users did not accept the ES recommendation in about 50% of the cases. The applicability of the modules Diagnosis and Treatment were above moderate as the advice of the ES fields supervisors matches the advice generated by the ES in the range of 80 to 87.5% in diagnosis and 50% of the cases in treatment. The applicability of the modules: Variety Selection, Irrigation, and Harvest are high. In the Variety Selection, the ES recommendations are compatible with the actual varieties cultivated in the ES fields. In the Irrigation, the Delta area (there is only 10% difference). In the Harvest, the ES recommendations are compatible with the actual practice in the ES fields.

7.3 Need of Expert System

In order to measure the need of NEPER, a comparison has been done between the advice given by the researchers and extension workers supervising the control field in the experiment locations and the advice that would be generated if NEPER were used. In the first experiment (CLAES, 1996), examining the comparison results it was found that the ES performance is better in 76% of the cases, and hence there is a great need for having the ES.

In the second experiment (CLAES, 1999), it was found that there is a high need for the ES modules: Tillage, Irrigation, Fertilization, Diagnosis, and Treatment. In the Tillage module, it was found that the performance of the ES is better as all control fields supervisors did not apply laser and plowers, appropriately. In the Irrigation module, it was found that the ES recommends less water than what was recorded in the control fields books. In the fertilization module, it was found that the ES is better as ES recommends the adequate quantities of phosphorus and potassium fertilizers whereas some control fields did not apply these types of fertilizers at all. In diagnosis and treatment modules, the performance of the ES is better as the advice of the control fields supervisors match the advice generated by the ES in only 37.5% of the diagnosis cases, and 20% of the cases in treatment. The experiment showed that there is a need for such module especially if the treatment part is modified to be more applicable.

8. EXPERT SYSTEM ENHANCEMENTS

According to the results obtained from field-testing, the following enhancements were done:

- Arabic language support was introduced.
- The irrigation module was revised to be accepted.
- User interface become more flexible.
- Basic information about the field and the enviromnent have been included in the reasoning (i.e. drainage system, previous crops, water source, length and width of the field, etc.)
- The variety selection module has been enhanced to produce the most suitable variety for each field and produce justification for this selection.
- Basin recommendation has been revised completely.
- The harvest module has been enhanced to generate real advice about the suitable date of start harvest

The following enhancements were also suggested and the ES are going to include them:

- Most of the users were unable to understand what was meant by some operations so, more explanation like video clips should be provided.
- Currently, the ES is capable of diagnosing sever nutrition deficiency. However, it is not equally capable of detecting early stages of nutrition deficiency. This should be rectified. A very good example of this is Nitrogen deficiency.
- Drought and Water Logging should be covered by the system specially that their symptoms coincide with the symptoms of Nitrogen and Potassium def.

9. CONCLUSION

The work done in this project has revealed and emphasized the effectiveness and importance of ES as a decision support tool for extension services. It was very clear that there is a difference in the advice quality and consistency given by the ES and the extension agriculture engineers.

In the mean time, field experiments showed that Usage of ES has an economic and environmental impact. Currently there are efforts to disseminate NEPER, nation wide, and to avail it on the Internet. The field testing was found to be very useful as many aspects of the usability, applicability, and need were not possible to be identified without this field test. NEPER was found to be user friendly, and can be used by both researchers and extension workers. The recommendations generated by NEPER were applicable in most of the cases. The cases that were not accepted by the researchers and extension workers conducting the experiment, were discussed and the right recommendations were included in the succesor version. Most of the NEPER modules are found to be needed. The modules which were found not needed , were examined. The result was that this was not needed by the researchers and extension workers conducting the experiment but they are badly needed by the growers and extension workers in remote locations.

REFERENCES

Adrion, W., Branstad, M., Cherniovsky, J.'Validation (1982) "Verification and Testing of Computer Software" ACM Computing Surveys, Vol. 14, No. 2

Chandrasekran, B. (1986). Generic Tasks in Knowledge-Based Reasoning: High-Level Building Blocks for expert system design. IEEE Expert, 1(3), 23-30.

CLAES (1996) "Validating NEPER Wheat Expert System and CERES Wheat Simulation Model", report, No: TR/CLAES/ATUT(1)/3/96.12, 1996.

CLAES (1999) "Validating NEPER Wheat Expert System - Field testing for season 97/98", report, No: TR/CLAES/ATUT(w4)/5/99.2, 1999.

CLAES (2001) "Validating NEPER Wheat Expert System - Field testing for season 98/99", report, No: TR/CLAES/ATUT(w4)/10/2001.3, 2001.

O' Keefe R.M (1990) "Consultant Report" Report No-CR-88-024-08 the Expert Systems for Improved crop management project No EGY/88/024

O' Keefe, R.M., O. Balci, and E. P. Smith, (1987) " Validating Expert System Performance " IEEE Expert, Vol. 2, No. 4, Winter 1987, PP 81-90.

O' Leary, D., O'Keefe, R. (1989) "Verifying and Validating Expert Systems", Tutorial: MP4, IJCAI,1989.

Rafea, A. (1996) "Natural Resources Conservation and Crop Management Expert Systems", Workshop on Decision Support Systems for Sustainable Development, UNU/IIST, Macau. 26 February - 8 March, 1996.

Gomez,F., & Chandrasekran, B. (1981). Knowledge Organization and Distribution for Medical Diagnosis. IEEE Transactions on Systems, Man, and Cybernetics, SMC-11(1), 34-42.

Kamel, A., Schroeder, K., Sticklen, J., Rafea,A., Salah,A., Schulthess,U., Ward, R. and Ritchie, J. (1995). Integrated Wheat Crop Management System Based on Generic Task Knowledge Based Systems and CERES Numerical Simulation. AI Applications 9(1):17- 27

A METHOD FOR DECIDING THE NUMBER OF HIDDEN NEURONS OF THE FEEDFORWARD NEURAL NETWORKS

Yuan Hongchun Xiong Fanlun Huai Xiaoyong
(Institute of Intelligent Machines, Chinese Academy of Sciences, Hefei 230031)
hcyuan@mail.iim.ac.cn flxiong@163.net

Abstract: The number of hidden neurons of the feed-forward neural networks is generally decided on the basis of experience. The method usually results in the lack or redundancy of hidden neurons, and causes the shortage of capacity for storing information or learning overmuch. This research proposes a new method for deciding the number of hidden neurons based on decision-tree algorithm. Firstly, an initial neural network with enough hidden neurons should be trained by a set of training samples. Second, the activation values of hidden neurons should be calculated by inputting the training samples that can be identified correctly by the trained neural network. Third, all kinds of partitions should be tried and its information gain should be calculated, and then a decision-tree for correctly dividing the whole sample space can be constructed. Finally, the important and related hidden neurons that are included in the tree can be found by searching the whole tree, and other redundant hidden neurons can be deleted. Thus, the number of hidden neurons can be decided. In the case of building a neural network with the best number of hidden units for tea quality evaluation, the proposed method is applied. And the result shows that the method is effective. *Copyright©2001 IFAC*

Keywords: neural network, decision-tree, hidden neurons

1. INTRODUCTION

Neural networks are widely used in many areas because of its strong capacity for non-linear mapping, its high accuracy for learning and its good robustness. A multi-layer feed forward neural network is a type of very important network, but it is very difficult to effectively decide the number of hidden neurons. At present, in many applications of networks, the main method to select the number of hidden neurons is as follows. Firstly, a series of neural networks with different number of hidden neurons are tested, then their inductive errors are evaluated, finally, one of them, which is the best one, is selected. This method cost much time and has great limitation. Some people propose that such experience as "the number of input neurons add the number of output neurons, then be divided by 2" could be taken to decide the number of hidden neurons. This method depended on the number of input neurons and output neurons is not scientific, because factors to affect networks' structure are the number of samples in a training set, the noisy size of samples and the complex degree of function or classification to learn, and so on. Some people also suggest that the maximal number of hidden neurons can be made according to the number of samples. This experiential method only can solve the problem of over-learning. Reference 1 puts forward a monotone index based method to directly estimate the number of hidden neurons in a three-layer feed forward network. This method requires a good set of sample set, and is not suitable to a small set of samples or multi-input and multi-output issue.

This research has proposed a novel method that the number of hidden neurons can be decided based on a decision tree algorithm. The goal of this research is to overcome above mentioned problems, and to avoid the over-learning problem because of over number of the hidden neurons, and to avoid the shortage of capacity because of less number of hidden neurons. Next parts will discuss the proposed method and present an example.

Supported by National Nature Science Fund of China (No.69835010)

2. PROBLEM DESCRIPTION AND ALGORITHM DESIGNATION

Problem description: if there is a training set S=<C,r>, $C=\{C_i|1\le i\le k, i\in N$, k=the number of condition attributes$\}, r\in R, R=\{R_j|1\le j\le m, j\in N$, m=the number of classes$\}$, then a three-layer feed forward neural network can be built to approximate the training data. The number of input neurons is k. The number of output neurons is decided on the basis of m. If m>2, then the number of output neurons is m. If m=2, then the number of output neurons is 1. How to estimate the number of hidden neurons is what this paper will discuss.

Toward above problem, this research proposes a method that using decision tree algorithm to decide the number of hidden neurons. The basic idea is as follows. Firstly construct an initial network with enough hidden neurons according to the number of training samples. Secondly train the network. Thirdly built up a decision tree, which can be used to divide the whole sample space based on the activation values of hidden neurons. And then search through the whole tree to find out important hidden neurons, which take part in correctly dividing sample space. Fourthly, take these neurons as hidden neurons of the final network. Thus the best number of hidden neurons can be made.

The algorithm, which decide the number of hidden neurons of network, can be described as follows:
Input: training set S=<C, r>
Output: the best number of hidden neurons of a neural network.
Algorithm steps:
Step 1: data preprocessing, such as filling missing data, standardizing data and so on.
Step 2: construct the topology structure of an initial network. The number of input neurons can be decided based on the number of condition attributes. The number of output neurons can be decided based on the number of classes. And the number of hidden neurons can be given enough according to the number of samples.
Step 3: initialize the original network.
Step 4: train the network until the error is less than predetermined value.
Step 5: select samples, which can be identified by the trained network, from the training set.
Step 6: calculate the activation value of the selected samples.
Step 7: sort all activation values of each hidden neuron, and make any possible partitions, then calculate their information gains to build a decision tree which can be used to divide sample space.
Step 8: search the decision tree to find out important hidden neurons, which are used in the tree.
Step 9: sum up the number of important hidden neurons, and take the result as the best number of hidden neurons.

Next parts will further discuss network training, decision tree constructing and important hidden neuron identification related to the above algorithm.

2.1 *Nnetwork Training*

Given a sample p, p=1,2,...,P, then the output value of the network and the activation values of the hidden neurons can be calculated as follows:

$$S_{ip}=\sigma(\sum_{j=1}^{J}V_{ij}H_{jp}-\theta_i) \quad (1)$$

$$H_{jp}=\sigma(W_jX_p)=\sigma(\sum_{k=1}^{K}W_{jk}X_{kp}-T_j) \quad (2)$$

Where, $x_{kp}\in[0,1]$ is the value of input neuron k of the given sample. w_{jk} is the connection weight from input neuron k to hidden neuron j. V_{ij} is the connection weight from hidden neuron j to output neuron i, T_j is the threshold value of hidden neuron j. θ_i is the threshold value of output neuron i. And $\sigma(\xi)=1/(1+e^{-\xi})$. J and K are the numbers of hidden neurons and input neurons respectively. Each sample x_p is one of possible classes such as $C_1, C_2, \dots, C_c$. t_{ip} can be marked as the teaching value of sample p in the output neuron i. Toward two-value classification issue, a neuron, which value is 1 or 0, can be used. Toward the classification issue of c>2, the number of output neurons can be c. If sample p belongs to c class, then $t_{cp}=1$, $t_{ip}=0(\forall i\neq c)$.

$$\theta(\omega,\upsilon)=F(\omega,\upsilon)-\sum_{i=1}^{C}\sum_{p=1}^{P}[t_{ip}\log s_{ip}+(1-t_{ip})\log(1-s_{ip})] \quad (3)$$

F(w,v) is an item of punishment:

$$F(\omega,\upsilon)=\varepsilon_1\sum_{j=1}^{J}(\sum_{i=1}^{C}\frac{\beta\upsilon_{ij}^2}{1+\beta\upsilon_{ij}^2}+\sum_{k=1}^{K}\frac{\beta\omega_{jk}^2}{1+\beta\omega_{jk}^2})$$

$$+\varepsilon_2\sum_{j=1}^{J}(\sum_{i=1}^{C}\upsilon_{ij}^2+\sum_{k=1}^{K}\omega_{jk}^2) \quad (4)$$

Where ε_1, ε_2 and β are positive parameters. Reference 2 shows that compared with standard least square error function, cross-entropy error function can be used to speed up the convergence. Reference 3 shows that punishment item can be used to accelerate the attenuation of weights. Thus, unimportant connections have small value. When it is less than a predetermined value, it can be set zero. By the method, the network connections can be simplified.

In the process of training the network, BFGS is used

to minimize the outstretched cross-entropy error function, because it can speed up the convergence compared with BP algorithm.

2.2 Decision Tree Constructing and Important Hidden Neuron Identification

After training a network, a decision tree can be constructed based on calculating the information gains of all partitions of hidden neurons' activation values, and related and important neurons can be identified by searching the decision rule of no-leave node.

If there is a data set D, then a method which produce a decision tree is as follows:

(1) If there are one or more samples in D, and all samples belong to C_c, then stop partitioning.

(2) If D is NULL, then take the most frequent class of its parent node as the class of this branch, and stop partitioning.

(3) If there are samples, which belong to different classes, then D should be partitioned based on the information gain.

This research adopts the activation values of hidden neurons to construct a decision tree, and it has following features: because each activation value is obtained from calculating sigmoid function, so all activation values are located in the continuous space (0,1). Because only using the activation values of hidden neurons of correctly identified samples to construct decision tree, so the number of samples in D generally is less than the total number of samples P. Let P' represents the number of samples which can be identified by the trained network. To hidden neuron j, the activation value of sample p can be represented as H_{jp} (p=1,2,...,P'). In the process of constructing a decision tree, firstly sort these values, and random divide them to two groups: $D_1=\{H_{j1},...,H_{jq}\}$ and $D_2=\{H_{j,q+1},...,H_{jp}\}$, and calculate standard information gains of all possible partitions among these activation values, then select the partition which has the maximal information gain from all partitions as the dependencies to construct a decision tree. For instance, if the standard information gain of hidden neuron j between mth sample and (m+1)th sample is maximal, and their activation values are H_{jm} and $H_{j,m+1}$ respectively, then the decision rule of root node can take $H_{ji}>(H_{jm}+H_{j,m+1})/2$, so that the activation values set of jth hidden neuron can be accordingly divided into two sub-sets, that means samples in D could be divided into left branch and right branch of the tree. The similar process can be adopted toward the two branches of root node, and such partition is continuous step by step, thus a complete decision tree can be produced.

The method to calculate standard information gain is as follows.

If there is a sample, which belongs to one of classes and exists in D, and n_c is the number of samples which belong to C_c, then its information entropy is:

$$I(D) = -\sum_{c=1}^{C} \frac{n_c}{N}\log_2\frac{n_c}{N} \qquad (5)$$

Where N is the number of samples in D. Toward the two branches of D, their information entropy can be calculated similarly:

$$I(D_1) = -\sum_{c=1}^{C} \frac{n_{c1}}{N_1}\log_2\frac{n_{c1}}{N_1} \qquad (6)$$

$$I(D_2) = -\sum_{c=1}^{C} \frac{n_{c2}}{N_2}\log_2\frac{n_{c2}}{N_2} \qquad (7)$$

Where n_{cj} is the number of samples which belong to C_c in D_j(j=1,2). $N_j = \sum_{c=1}^{C} n_{cj}$. The information gain of dividing D into D_1 and D_2 is:

$$Gain(H_{jq}) = I(D) - [I(D_1) + I(D_2)] \qquad (8)$$

The information gain can be standardized as follows:

$$NGain(H_{jq}) =$$

$$Gain(H_{jq})/[-\sum_{j=1}^{2}(N_j/N)\log_2(N_j/N)] \qquad (9)$$

Once a decision tree has been constructed, it is easy to identify the related and important hidden neurons. The important hidden neurons are ones that are applied in the nodes of decision tree.

3.EXAMPLE ANALYSIS

This section will take building a neural network model with the best number of hidden neurons for reflecting the relation between tea quality and quality effecting factors as an example to introduce the application of the proposed method.

Reference [6] presents that tea quality is mainly determined by such eight factors as three kinds of chromacity values (L*, a*, b*), four kinds of chemical compositions and unit weight. The data of this example is selected from reference 6. The total number is 36. Each sample has above eight attributes and one class attribute. The quality of CHAOQING green tea is usually expressed as grades, and the total number of grades is 7. So the topology structure of the initial neural network is as follows. The number of input neurons is 8, the number of output neurons is 7, and the number of hidden neurons can be given 10 by estimation and it is greater than it should be.

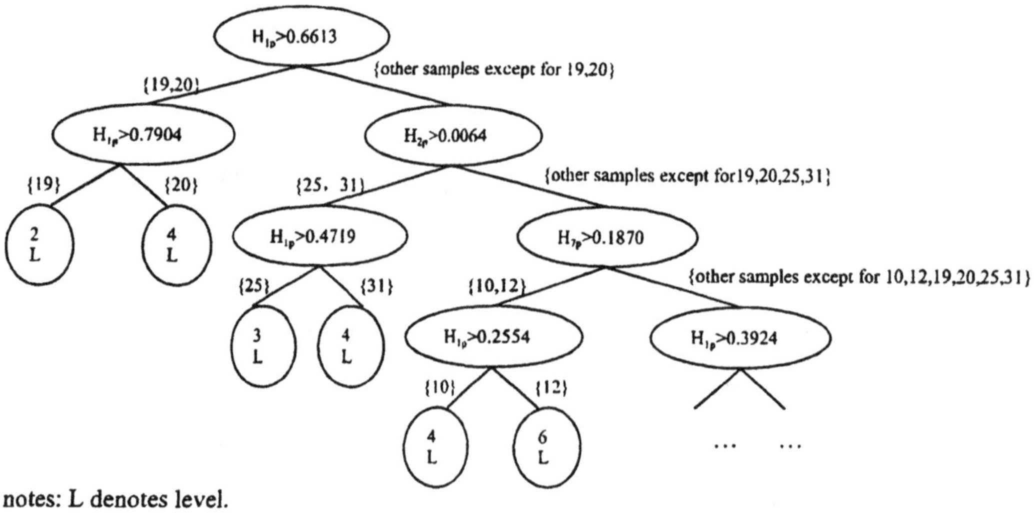

notes: L denotes level.

Fig. 1. a decision tree that can correctly divide samples space

Table 1 the relation between the number of hidden neurons and learning times

NUM_HID / ID	5	6	7	8	9	10
1	6173	4354	3430	6593	4106	7251
2	7348	7490	7197	4724	3905	6766
3	4469	6296	7409	4163	4483	8116
4	6990	9954	4235	8369	11021	7233
5	18260	11684	4788	3224	7807	3597
average	8648	7956	5412	5415	6264	6593

Notes: NUM_HID stands for the number of hidden neurons, ID stands for the serial number.

The above neural network should be initialized, and its outstretched cross-entropy error function can be minimized by BFGS algorithm. When the network is convergent, the trained network can identify 36 samples correctly, so each hidden neuron has 36 activation values. Then sort these activation values, random partition them, and calculate its information gain, finally a decision tree, which correctly divides the samples space, can be produced according to the above mentioned algorithm (see Figure 1). H_{ip} in the nodes of the tree represents the activation value of the pth sample in the ith hidden neuron. Through searching the tree, 31 non-leave nodes are included in the tree, and 7 related and important hidden neurons (1,2,3,5,6,7,9) are involved. So in this case, the best number of hidden neurons is 7.

To testify the effectiveness of this result, try method is adopted. That is, different neural networks with different hidden neurons are tried, and find out the best one that its learning times is least when what the number of hidden neurons is. To overcome stochastic error, 5 times experiments have been conducted. In the process of experiments, 0.01 is adopted as the convergent threshold. The result of experiments is shown as table1.

From table 1, it can be easily found that the best number of hidden neurons is 7, and it accords with the result, which is obtained by the proposed method.

4.CONCLUSION

This paper proposes a novel method to estimate the best number of hidden neurons of a network by firstly constructing a decision tree based on calculating information gains of all partitions among hidden neurons' activation values, and then searching the decision rule of the tree's nodes to find out important hidden neurons. Related algorithm and an example are presented in the paper. Result shows that the proposed is effective. This approach is latently significant for neural networks to approximate training samples, for researching rule extraction from neural networks and for implementing neural networks using hardware.

REFERENCES

LI yujian (1999).

a method to directly estimate the number of the hidden neurons in the feed forward neural networks. Journal of computers(in chinese), **Vol.22.No.11**, 1204-1208.

A.van Ooyen, A.and B. Nienhuis, B (1992). *Improving the convergence of the backpropagation algorithm*. Neural Networks, **vol.5, no.3**, PP.465-471.

J.Hertz,A.Krogh,and R.G. Palmer (1991). *Introduction to the theory of neural computation*. Redwood City, CA:Addison Wesley.

R.Setiono(1995). *A neural network construction algorithm which maximizes the likelihood function*, Connection Science,**vol.7,no.2**,pp.147-166.

R.Battiti(1992). *First- and second-order methods for learning: Between steepest descent and newton's method*. Neural Computation, **Vol.4**, PP.141-166.

LinGang(1991). *using multi-variable analysis method to evaluate the quality of Chinese tea and Japanese tea*. doctoral thesis of MingCheng University.

NEURAL NETWORK BASED FAULT DETECTION IN HYDROPONICS

K. P. Ferentinos[1], L. D. Albright[1], B. Selman[2]

[1]Department of Agricultural and Biological Engineering
[2]Department of Computer Science
Cornell University, Ithaca, NY 14853, U.S.A.

Abstract: A fault detection model for hydroponic systems, based on the feedforward Neural Network methodology was developed. Three kinds of faults were considered: mechanical, sensor and biological faults. In this paper, a preliminary detection system is presented, which generally detects the existence of any faulty situations. In the developed network, only the two first kinds of faults were considered. Biological faults, because of their particularities, were treated separately and some of their characteristics are presented at the end. *Copyright © 2001 IFAC*

Keywords: Neural Networks, Fault Detection, Hydroponics, Backpropagation Algorithms.

1. INTRODUCTION

The goal of every greenhouse facility is to maximize quantity as well as quality of production. This maximization is achieved using automated systems in order to control the environment inside the greenhouse, so optimal conditions for the specific cultivated plant are approximated. In addition, hydroponic cultivation gives the opportunity to control root environment precisely and thus to have more extensive plant production system control.

Loss of control over a greenhouse is a phenomenon usually having negative effects on the production and, consequently, the profit of the facility. Certain failures (faults) are easily noticed, but others are quite difficult to detect. Of course, a feedback-controlled greenhouse may be able to maintain desired conditions even when some parts of the control mechanism are out of order. However, especially when faults concern the plant, the effects may be disastrous for the entire production. Thus, detection and diagnosis of possible faults becomes very important. In particular, that which is the most important is fast detection of incipient faults. That is, detection at the earliest possible stage of slowly developing faults, as well as the quick identification of the problem. In order to have quick detection, an on-line identification system is necessary.

A feedforward Neural Network (NN) based fault detection system was developed. The main area of concentration was deep-trough hydroponic systems. This becomes easier in hydroponics, for several environmental variables of the plant can readily be monitored. The cultivated plant was lettuce (*Lactuca sativa*, cv. Vivaldi). The main variables of the root environment that can be monitored and can give an image of the real situation of the system are the nutrient solution's pH, Electrical Conductivity (EC), dissolved oxygen (DO) as well as its temperature and in addition, the transpiration rate of the plants.

2. MATERIALS AND METHODS

Behaviors of normal and malfunctioning systems may differ completely and those differences may be detected by measuring the environmental variables mentioned before. The main advantage of the use of Neural Networks is that the exact relation between the values of those environmental variables and the situation of the system (normal of faulty), is not needed. This means that it is not required to have an accurate model of the hydroponic system, from which several residuals or fault signatures would be extracted and used for the final fault detection. Neural Networks have been proved capable of identifying faults in several complex biological processes (Parlos *et al.*, 1994; Sorsa *et al.*, 1991; Venkatasubramanian and Chan, 1989; Watanabe *et al.*, 1989) in which, neither analytical models nor intermediate residual calculations were used.

Lettuce plants were cultivated in a continuous production deep-trough hydroponic system. They were transplanted into the system after growing from seed in a growth chamber for 11 days. Every two days, seven new plants were transplanted to the system, while seven plants of the age of 28 days (from seeding) were harvested. Consequently, the three ponds of the system had constantly the same number of plants of the same age. This made the system somewhat stable or, more precisely, periodically stable, with a period of two days. The important advantage of this method, except for having a quasi-stable system, is that a continuous production of plants was achieved (harvest every two days), which resembles real-life hydroponic production systems more closely than other techniques previously used in neural network modeling of hydroponics (Ferentinos, 1999).

Desired values of the environmental parameters during the operation of the hydroponic system were an air temperature of 24° C during the day and 19° C during the night, relative humidity from 30% to 70% and a light integral of 17 $moles \cdot m^{-2} \cdot d^{-1}$. For the nutrient solution, the pH set point was 5.8, the EC was maintained between 1150 and 1250 $microS \cdot cm^{-1}$ and the DO was maintained between 6 and 7.5 *mg/L*.

2.1 Normal and Faulty Situations

The procedure of training the NN requires an accurate definition of "normal operation", defined in our case as unstressed plants in a system that is in control. We need not express this normal situation by means of specific values of the environmental variables, because neural networks do not need such a representation in order to learn the pattern. Thus, we need to know only when the system and, in extension, the plants are in conditions considered to be normal by the producers, and also to know which training data sets correspond to those normal conditions. The values of the measured variables mentioned before were considered to be the normal values for the system and the plants were considered to be normal as long as they appeared healthy. We also need to define the "faulty operation" and to categorize this kind of operation into different types of faulty operations, one for each different kind of fault. In order to take data sets for each kind of fault, we have to impose those faults and take the corresponding measurements of the microenvironment variables.

Because the NN was trained off-line (meaning that the data sets were first collected and then used for training), there was no way of mistakenly having unhealthy plants in data sets of "normal operation". The "faulty operation" consisted of three different kinds of faults:

- Faults in actuators of the hydroponic system,
- Faults in sensors of the hydroponic system and
- Faults in the plants themselves.

More specifically, the faults considered, by category, are the following:

Mechanical faults. These are failures in mechanical parts of the hydroponic system, such as: a) pH control pump is out of order, or b) circulation pump is out of order.
Sensor faults. The ones considered were: a) pH sensor failure and b) EC sensor failure.
Biological faults. These are problems in the cultivated plants themselves and are divided into: a) root area faults and b) shoot area faults.

For the first two categories of faulty operations, real data exist, as a lot of sensor and actuator failures were encountered during the set-up of the system. In addition, several faults were especially imposed in order to train the NN model and investigate its inherent fault detection capabilities. For the third category however, faults were imposed directly on the plants. To cause or to try to imitate the effects of a possible fault in the root zone, the plants were removed from the ponds and the root were exposed to air for intervals of five minutes. In the case of imitating a modest fault in the shoot area, leaves were disturbed (mechanically) for intervals of five minutes and slightly damaged in doing so. Finally, to imitate more permanent damage to the plants, several experiments were performed by cutting several leaves of each plant or by covering each plant with transparent plastic bags.

2.2 Neural Network Fault Detection Model

The feedforward methodology of neural networks was used. The inputs of the NN were the environmental parameters (air temperature, relative humidity and light intensity), the measurable variables of the microenvironment of the plant (pH, EC, DO, nutrient solution temperature and transpiration rate of the plants) and the control signals of the pH and the DO control schemes (amounts of acid and oxygen added, respectively). Each output of the NN corresponded to a specific fault and there was also one output that corresponded to normal operation.

Several different architectures of one-hidden-layer and two-hidden-layer networks were tested, with two different activation functions (logistic and hyperbolic tangent). A new methodology for optimal network design and parameterization, based on Genetic Algorithms, was developed, but its results are incomplete, so only results of the conventional approach are presented. The training methodology was the Backpropagation Training Algorithm (Rumelhart *et al.*, 1986). Four different multidimensional minimization algorithms (steepest descent, conjugate gradient, quasi-Newton and Levenberg-Marquardt algorithm) were tested. An on-line adjustable learning rate performed better than a constant one. In the steepest descent and the conjugate gradient algorithms the Hessian was used at every iteration to solve for the "best" learning rate. For the other two algorithms, the "best" learning rate was calculated with an approximate line search using a cubic interpolation. The final NN fault detection system was tested using new data and its generalization capabilities were explored.

In addition to the network inputs listed before, one-step and two-step histories of the pH, EC and DO variables were included. That is, for each of these variables, three inputs existed: one for time t (current time), one for time t-1 (previous time step) and one for time t-2 (two time steps before). Thus, the network had 15 inputs. The time step was 20 minutes.

The final neural network is going to have one output for the normal operation and an output for each of the faults considered. However, the amount of data collected so far is not yet sufficient to train such a network and in addition to test its performance. Therefore, a simpler neural network was trained and tested as a preliminary step. This NN had all the inputs presented before except for the transpiration rate, but only one output, a binary output having the value zero corresponding to normal operation and one corresponding to faulty operation.

3. RESULTS

In this paper, a preliminary investigation of the performance of the general normal/faulty operation detection neural network is presented. All mechanical and sensor faults of the systems were treated as a general "faulty situation". Biological faults showed no correlation with any measured variable except for the transpiration rate. On the other hand, transpiration rate was not affected by the first two fault categories. Therefore, biological faults were not considered in this preliminary model and in addition, transpiration rate was not used as an input to the network. The following section describes the training and evaluation of the preliminary NN fault detection system, while the next section analyzes some first results of some imposed biological faults.

3.1 Neural Network Performance

The NN model was trained with experimental data collected from all three ponds of the system, for both normal and faulty situations. Approximately 15 (non continuous) days of data of both normal and faulty operations formed the final training set. The time step of these data was 20 minutes. Thus, the NN training was based on 1050 entries for each input.

The training process included two basic parts. The first part, the preliminary training process, determined the best combination of network architecture and training algorithm. This was achieved by training several candidate network topologies (both 1-HL and 2-HL networks) with all four training algorithms and comparing the results. The second part of the training process, which was the *basic training process*, focused on training the best combination of architecture and algorithm. Based on the preliminary training, the network architecture/algorithm combination that gave the best results was the 2-HL NN with 9 nodes in the first hidden layer and 9 nodes in the second hidden layer, trained with the quasi-Newton algorithm.

The basic training process had a goal to further train the selected NN with the best possible algorithm for this system and architecture, which was proven to be the quasi-Newton algorithm. Many different random initial network parameters were tested in that training. Also, several values of the coefficient of the penalty term for the regularization (λ), varying by the order of 5, were tried. Both logistic and hyperbolic tangent activation functions (functions of hidden nodes) were tested. The results of these tests showed that the value of λ that leads to the minimum sum squared error (SSE) was λ= 0.05 and that the network with logistic activation functions performed better than the one with hyperbolic tangent activation functions. Thus, the final NN model consisted of a network with 15 inputs, two hidden layers with 9 nodes each that have logistic activation functions, and one output. The preferred training algorithm was the backpropagation training algorithm using the quasi-Newton multidimensional minimization algorithm with parameter λ=0.05.

Testing consisted of presenting new data to the trained NN model and exploring its generalization capabilities. The main goal in a fault detection system is not only detection of the existence of a fault, but also its rapid detection. Especially, when we deal with incipient faults, the time factor becomes more important because these faults are more difficult to detect as they begin. Six testing data sets were presented to the NN fault detection system. Each set starts with data of normal operation and some specific fault is imposed at some known time, except for the last data set that contains only normal data.

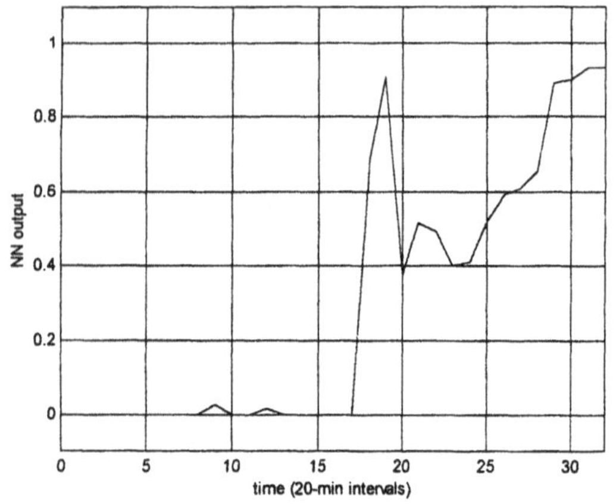

Fig. 1. Output of the FDNN for testing data set 1 (fault imposed at the 16th interval).

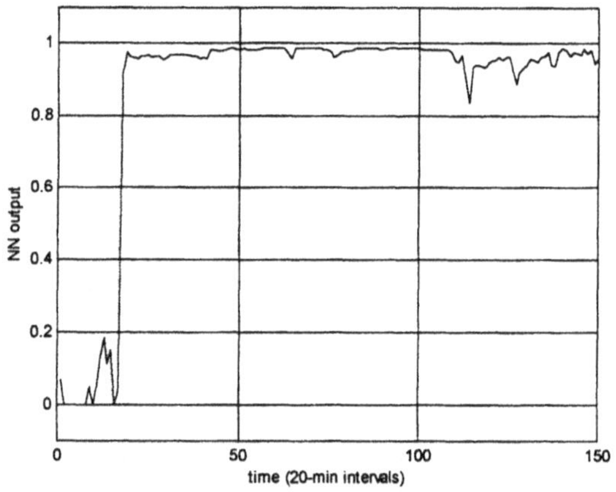

Fig. 2. Output of the FDNN for testing data set 2 (fault imposed at the 18th interval).

The output of the NN was considered to represent a faulty operation if it has a value greater than 0.5, while for values smaller than 0.5 a normal operation was assumed. Fig. 1 shows the NN output on the first data set. The "pH control pump out of order" fault was introduced at the 16th interval. The fault was detected by the network within only two step intervals (point 18), that is a period of 40 minutes. After that the network gives a steady and strong indication that the operation is not normal, with values very close to 1.

Fig. 2 shows the network output for the second testing set, which contains a "circulation pump out of order" fault. The data again start with normal operation and the fault is introduced in the 18th interval. It is clear here that this kind of fault is detected very fast. Even from the 18th data point, the output of the network is 0.69, which is considered as a weak fault indication. The next data point gives a value of 0.91 that strongly indicates the existence of faulty operation. A disadvantage here can be considered the fact that the output drops below 0.5 (normal operation) 40 minutes after the occurrence of the fault and stays in that area for an hour. After that period it returns to the faulty indication. This is not something important, if we consider the fact that the indication of some fault is already given at a very early stage and also that the kinds of faults considered are supposed to be irreversible without the interaction of a human factor with the system.

In fig. 3, the network response in the third testing set is presented. This set contains the same fault type as the previous one that is now introduced in the 9th data point. This time it takes 20 minutes to the network to indicate a possible fault (value 0.64) while in a total of one-hour period after the introduction of the fault, the output becomes high enough (0.82) to strongly indicate the existence of faulty operation.

The fourth testing data set has 167 20-minute intervals and the "failure in pH sensor" fault was introduced in the 16th point. The output of the NN model is presented in fig. 4. Similarly to the previous case, it takes one time step (20 minutes) for the network to indicate a possible fault with an output of 0.60, while at the next 20-minute step the output becomes 0.91 and stays in that area. The rather periodical fluctuations of the output are caused by the nature of the sensor fault. This kind of fault was reproduced by adding a periodically changing noise to the readings of the pH sensor. The form of noise is

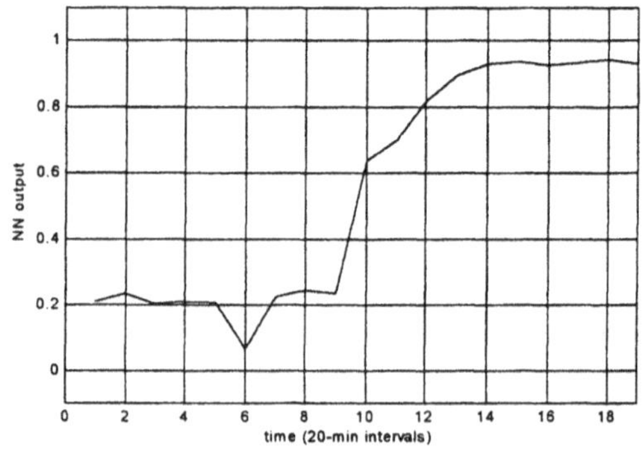

Fig. 3. Output of the FDNN for testing data set 3 (fault imposed at the 9th interval).

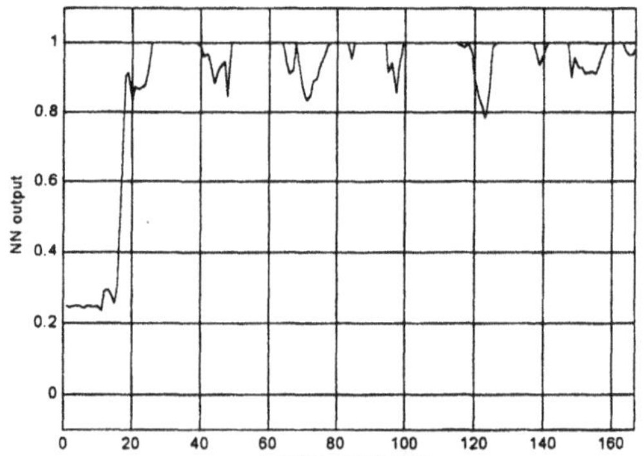

Fig. 4. Output of the FDNN for testing data set 4 (fault imposed at the 16th interval).

a sine function. Thus, in the points where the noise becomes small, the fault confidence of the network decreases. However, this does not cause the network to exit the "faulty situation" area.

The fifth testing data set contains the implementation of the "failure in EC sensor" fault. The set has 69 20-minute intervals and the fault was introduced in the 16th interval. As can be seen in fig. 5, the fault is detected 4 hours after its beginning. Moreover, several hours later, when the noise of the sensor failure becomes small, the network indicates normal operation for that period. It seems that this specific fault causes some problems for the detection process, probably because no information about the control signal for the EC is present; EC was controlled manually in the hydroponic system.

Finally, the last testing data set contains only normal operation data and it was used to check the network ability to recognize continuously varying normal behavior. The output of the network is shown in fig. 6. The graph shows a period of almost one whole day and the output is always below 0.4, indicating normal operation during the entire period.

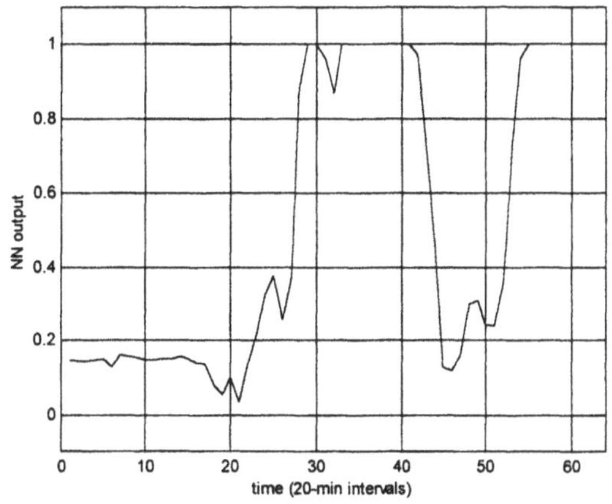

Fig. 5. Output of the FDNN for testing data set 5 (fault imposed at the 16th interval).

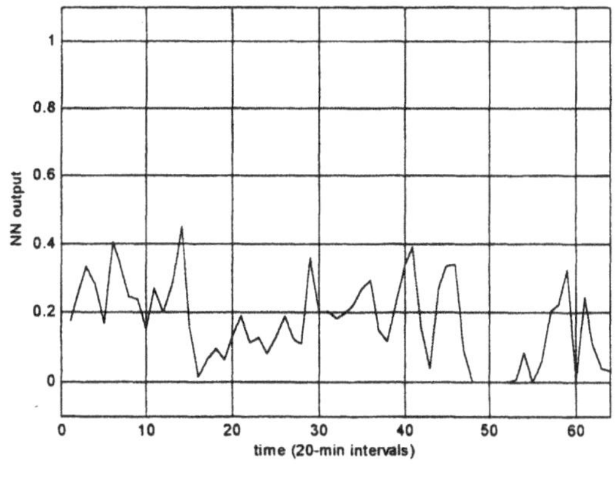

Fig. 6. Output of the FDNN for testing data set 6 (normal operation).

3.2 Biological Faults

Several experiments consisted of imposing different faults on the plants in order to examine their effect on the monitored variables of the nutrient solution of the system. Four different series of experiments were performed. In the first, most of the largest plants were removed from the pond for five minutes. In this way, possible problems in the root zone of the plants were imitated. In the second series, several leaves of the largest plants were removed. This action imposed permanent damage on the plants and imitated the effects of major problems in the shoot zone of the plants. A similar but less influential series of experiments was the one in which leaves of the plants were disturbed for intervals of five minutes and slightly damaged. These experiments imitated shoot problems less important than the ones imitated in the previous series of experiments. Finally, in the fourth series, the largest plants (of ages of 23, 25 and 27 days) were covered with transparent plastic bags. This imposed a temporary fault that imitated minor problems in the shoot zone.

Effects of these biological faults, unfortunately, were not significant enough to be used in a fault detection scheme. The pH and the electrical conductivity appeared not to be affected at all by the faults. The transpiration, a variable known to be drastically affected by the condition of the plants, was so highly correlated with the environmental conditions of the greenhouse (temperature, light intensity and relative humidity) that effects of plant damage, even when seemingly severe, were not noticeable in most of the cases. Even in the experiments in which some leaves of the plants were cut and where one would expect major impacts in the transpiration ratio, the effects were "hidden" by the high correlation of the transpiration with the environmental parameters, especially temperature and light intensity. In fig. 7, the differences between the cumulated evapotranspiration rates between two of the tanks of the hydroponic system are shown. Every set of points represents periods of two days, between transplanting. At those points, the cumulated differences were reinitialized. At around time interval No. 2000, half of the leaves of the largest plants of tank 1 were removed, while nothing changed in tank 2. One would expect that the transpiration would be reduced in tank 1, thus the difference between transpiration rates of tank 2 and tank 1 would increase. As can be seen in the graph, this is clearly not the case. This can be explained if the additional transpiration from the cuts of the removed leaves is taken into account. Thus, no indication of the fault appeared in this case. The differences between tanks even when normal conditions exist in both tanks, is caused by differences in the air movement above the tanks, which lead to difference transpiration rates. Similar results were obtained by the imposed faults of disturbing the leaves or removing the largest plants from the tanks for periods of five minutes.

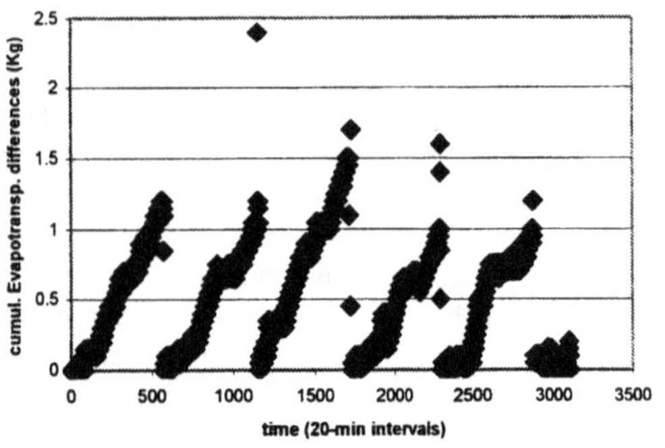

Fig. 7. Cumulative evapotranspiration differences between two tanks of the hydroponic system ("cutting the leaves" fault imposed at the 2000th interval).

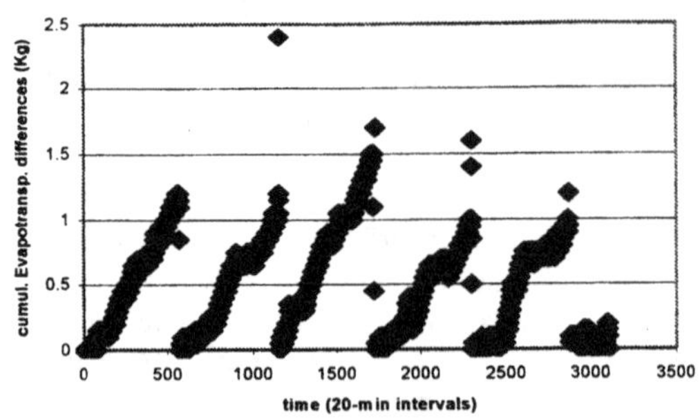

Fig. 8. Cumulative evapotranspiration differences between two tanks of the hydroponic system ("covering the plants" fault imposed at the 1000th interval).

The only biological fault that gave an indication of its existence in the rate of transpiration was the fourth type, during which the largest plants were covered with transparent plastic bags. As shown in fig. 8, significant increase in the difference of cumulated transpiration rates between tank 2 and tank 1 occurred after the introduction of the fault in tank 1 (at time interval No. 1000). It is clear that during transpiration rate was reduced in tank 1 (the difference from transpiration of normal plants increased) during the six following to the introduction of the fault days. During these days, covered plants existed in the tank. After that period, all covered plants had been harvested, thus transpiration rate in tank 1 returned back to normal (last two sets of points in the graph).

More data of this type of biological faults have to be collected so that a detection system can be developed. Because of the large amount of history data of the transpiration of the system that is needed by such a detection scheme, this model is going to be separate from the NN model the detects the other kinds of faults (mechanical and sensor faults).

4. CONCLUSIONS

The methodology of constructing a neural network based fault detection system in hydroponics was developed. Some preliminary results are presented. The results reflect a simplification of the more general NN model in that it has only one output and tries to classify specific data into either normal or faulty operation. These testing results indicate that this simplified network is capable of detecting the faulty situation is very short time in most cases. The rapidity of detection suggests time steps smaller than the 20-minute time step used here. The results show that the NN has useful generalization capabilities. A next step is to develop a more detailed fault detection system that has one output for each specific fault.

ACKNOWLEDGEMENTS

This research was supported by BARD Research Project No. IS-2680-96.

REFERENCES:

Ferentinos, K.P. (1999). *Artificial Neural Network Modeling of pH and Electrical Conductivity in Hydroponic Systems.* MS Thesis, Cornell University, Ithaca, NY.

Parlos, A. G., J. Muthusami, and A. F. Atiya. (1994). Incipient Fault Detection and Identification in Process Systems Using Accelerated Neural Network Learning. *Nuclear Technology*, **vol. 105**, pp. 145-161.

Rumelhart, D. E., G. E. Hinton, and R. J. Williams. (1986). Learning Representations by Back-Propagating Errors. *Nature*, **vol. 323**, pp. 533-536.

Sorsa, T., H. N. Koivo, and H. Koivisto. (1991). Neural Networks in Process Fault Diagnosis. *IEEE Transactions on Systems, Man, and Cybernetics*, **vol. 21**, No. 4, pp. 815-825.

Venkatasubramanian, V. and K. Chan. (1989). A Neural Network Methodology for Process Fault Diagnosis. *AIChE Journal*, **vol. 35**, pp. 1993-2002.

Watanabe, K., I. Matsuura, M. Abe, M. Kubota, and D. M. Himmelblau. (1989). Incipient Fault Diagnosis of Chemical Processes via Artificial Neural Networks. *AIChE Journal*, **vol. 35**, pp. 1803-1812.

NEURAL NETWORKS FOR INFLUENCE ANALYSIS ON THE QUALITY OF POTATOES

Sascha Richter[1], Klaus Gottschalk[1], Erhard Konrad[2]

[1] *Institut für Agrartechnik Potsdam-Bornim,*
Max-Eyth Allee 100, D-14469 Potsdam
[2] *Technische Universität Berlin,*
Franklinstraße 28/29, D-10587 Berlin, Germany

Abstract: A model to predict the occurrence probability of blackspots on potatoes is presented. The assessment of the quality factors in point of view of their usability as input into the prediction model, based on a competitive approach, is discussed. Neural networks are distributed in a computer cluster. These neural networks are trained with pre-recorded datasets stored in an online database. Variants of neural networks are created, differing in their preprocessing capability of the input vector. The results of the prediction system are compared with the data from real blackspot indices of measured surfaces of potato samples. The prediction error is re-input into the neural network for backpropagation, and the prediction quality of the network is improved. The process for calculating different variants of neural networks at the same time is shared of several workstations in a cluster. *Copyright © 2001 IFAC*

Keywords: Neural networks, Knowledge based system, Computer networks, Database systems, Potato quality

1. PREDICTION MODEL BASED ON NEURAL NETWORKS

One of important lacks of quality on potatoes is the occurrence of blackspots. Blackspots can be measured only with destructive methods. A prediction model may help to reduce these destructive tests. Many influencing factors of the occurrence of blackspots are well known, other are unknown, see Wormanns et al. (1997). At the Institute for Agricultural Engineering Potsdam-Bornim a method based on neural networks was developed by Richter et al. (2000). The advantage of using neural networks are their selflearning capabilities, and the fact that they are able to handle fuzzy knowlegde, see for example (Simon 1994) and Brause et al. (1995). Neural networks are used here to predict the occurrence of blackspots after a training process.

The ingredients of the potatoes and other influencing factors like mechanical stress, temperature, and others are analysed. The results of this analysis are stored together with the measurements of the blackspoted surface of the potato in a database. Since 1995, charges of 240 potatoes of the variety Adretta, Likaria, and Koretta are analysed. The resulting database is accessible by the internet with a web frontend. The recorded influencing factors are the input vector to the neural network. Because that every blackspot on the analysed surface of every potato is recorded, this value is used as the target vector of the neural network. A dataset of the influencing factors is fed into the neural network and the output vector is compared against the real measured blackspoted area. The difference between the real size and the predicted size is the error of the neural network. This difference is re-fed into the neural network to adjust the internal thresholds and weight vectors of the neural network. After this correction, the output of a neural network should be closer to the real size of the blackspots occurance. It is obvious, that the used neural network is a kind of a backpropagate neural network.

The quality of a neural netwok depends on the correct pre-processing of the input vector because input vectors are reflecting in a linear way to the output vector. The preprocessing of the input vector is limited to four operands (^-2, ^-1, ^1, and ^2), where ,^' means ,power of...', resulting in 16 different variants of neural networks. Each of these variants gives an independent neural network which has to be trained seperately.

Some of the observed influencing factors may have a high influence on the quality of the prediction, other factors may have low or no influence. The importance of the prediction rate is assessed and given by a quality number resp. quality index.

Neural networks (NN) with a good prediction rate get a higher quality number than neural networks with a bad prediction rate. Factors without influence to the occurrence of blackspots result in a neural network with a bad prediction rate. Every combination of an influencing factor with another is used as the input vector of a NN Every of the 16 variants of NN is evaluated. The NN is trained and according to its prediction rate, the quality number is assigned in to the database of the used influencing factors. At the end of the evaluation, it is obvious, that influencing factors with a high quality number had been best performed in the sense of their usage as input vector, resulting in a NN with a good prediction rate.

Prediction systems based on neuronal networks are trained with a sample set of potatoes. The test of the neuronal network trained with the sample set data shows a good prediction rate, but shows a significant lower prediction rate with an untrained new charge. Figure 1 shows a neuronal network trained with the charge 1 1995. The input vector involves the mechanical stress index and the temperature as influencing factors. The neuronal network is trained to separate potatoes into two classes with less than 2% and greater than 2% of blackspots on the surface. A good prediction system based on neuronal networks needs to be retrained with samples of the tested charge to give a high prediction rate. The process of calculation of all combinations of neuronal networks needs a high performance workstation, because the calculations have to be repeated with new samples and also for every variety of an agricultural product like potatoes to reach a high level of prediction rate.

1.1Assement of three different varities

A neural network has to be retrained to keep its level of prediction rate as it is shown in figure 1. The prediction rate is significantly higher after a retraining of the neural network with test samples of a charge for assessment. This leads to the conclusion, that this prediction model has to be adjusted to every charge to give the best results. A general prediction model will probably never reach an accuracy of more than 70%.

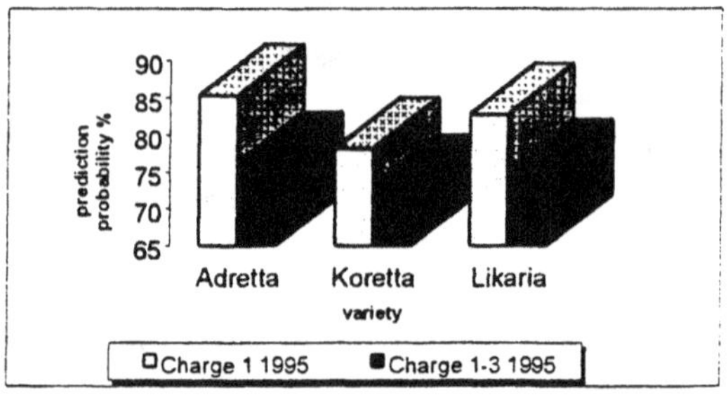

Fig. 1. Prediction system to classify the occurrence of blackspots on potatoes trained with the influencing factors mechaincal stress and temperature

1.2 The concept of the cluster

The easiest way to increase the computational performance is the application of a cluster of workstations . The classic Beowulf cluster concept based on programming libraries like PVM or MPI described in ANTLR (anonymous), are not the adequate solution to this problem, because these libraries use low-level synchronisation and communication.

A specialist cluster concept integrating internet technologies and a new object-model meets the requirement for a system to calculate several neuronal networks. A cluster application is distributed over several workstations. This concept makes it necessary that the workstations communicate with each other. On a Local Area Network (LAN) the communication is normally not restricted. But most LANs are protected by a firewall, which restricts the communication on some protocols like HTTP, HTTPS or EMAIL.

Workstations in the clusters have one of the following roles:

- The **algorithm server** handles the compilation of the source code and the transfer of the resulting binary programs to the workstations. It co-ordinates the synchronisation and the communication.
- The **algorithm client** executes the binary programs

received from the algorithm server. Also, the results are sent back to the algorithm server.

- The **interface server** is basically a web server. It sends forms based on the Hypertext Markup Language (HTML) and Java-applets to the interface clients.
- The **interface client** is a Java enabled web browser. The cluster is managed by HTML forms and applets. The application framework is downloaded once, when the application is executed for the first time. When Java Webstart is already installed, no additional software is needed.

Interface clients may be located somewhere in this world. The workstation acts as an interface client and uses one of the allowed communication protocols.

The most preferable protocol is the Hypertext Transfer Protocol (HTTP), because the client software is already installed on most workstations in form of a Java enabled web browser and most firewalls accept communication over HTTP.

This means, that in the most cases an installation of a software on the client side is not necessary and the costs of support are minimised.

The interface and algorithm servers are installed on the same machine, but this is not a necessarily required.

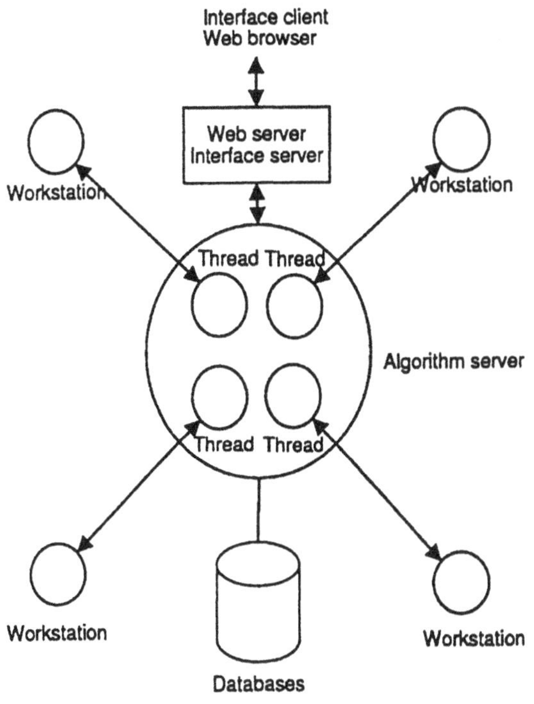

Fig. 2. Structure of the cluster

The communication between the algorithm server and the algorithm clients are based on CORBA. CORBA is an object model which makes it possible to use functions written for one programming language in an other programming language. CORBA makes it easy to access remote functions. The algorithm clients may run on different operating systems. CORBA has two more advantages:

- free available implementations (MICO, JacOrb etc.)
- support for multi-threading

The support of multi-threading in the object model is crucial to the success of cluster operating systems. The algorithm server is a program with several threads. Each thread communicates with one workstation in the cluster. The advantage of threads is that they share the same program context, like variables, file descriptors, etc. A program may start and stop threads and has full control over these threads. The threads communicate with the main program by semaphores, function calls or exceptions.

Performance-critical parts of the cluster were written in the programming language C++. The code for the communication interface used on the server-side and the client-side is written in the programming language Java.

The programming language in which software for the cluster are developed are written in the programming language "Prolog". The standard Prolog has no support for clustering, CORBA or communication over TCP/IP. The required programming language has to be developed as a bottom up procedure.

The algorithm server is the center of the cluster. The algorithm server is responsible:

- for the synchronisation of the base cluster software
- for the scheduling of algorithms to the algorithm clients
- for the compilation of the user programs
- to run control of the user programs

1.3 The synchronisation of the base cluster software

The framework of the cluster, is the software that transfers the user programs to the workstations in the cluster, synchronises the user programs, and handles the communications. This software is the operating system of the cluster. A cluster is a distributed software, every workstation is running this cluster operating software. But the cluster software is like most software products: a product in development. The interfaces, the structures, the classes, etc. may change. A workstation running an older cluster operating system than an other workstation may not be able to communicate correctly with this workstation.

Therefore, it is absolutely necessary, that each member of a cluster is running the same version of the cluster software.

The algorithm server is checking whether the cluster operating System is up to date on all workstations in the cluster with a low level protocol before every cluster session. Older software components are upgraded. This process happens every time a client connects or reconnect to the algorithm server. The upgrades of the software libraries are hidden to the users of the clusters. They have no influence on this process.

1.4 The scheduling of the algorithms

The users write their source code with the tools offered by the interface server. From these source code the user program is created. The algorithm server is responsible for the transfer of the user programs to the workstations in the cluster. The user programs are separated in parts. Has a workstation terminated his program, it sends the results to the algorithm server. The algorithm server sends a new part of program to the workstation for calculation. The scheduling of algorithms is the most import part of the cluster software, because most of the performance issues are dependent on how clever the mechanism of algorithm scheduling works. The user is able to influence the distribution of the application in the cluster. The keyword ***single*** limits the followig function to the distribution to one workstation only. The modifier ***server*** limits the following function to the execution on the application server. The modifier ***multible:X*** constrains the involved workstations to the maximum of X, where X is a natural number. The idea behind these keywords are simple. Some algorithms may not be calculated in smaller tasks or the synchronisation costs are higher than the resulting performance improvement by a distributed calculating. Some functions should only run on the server, because they are responsilbe for specialysed synchronisation tasks. With these keywords a heuristic optimization may be achived at design time.

1.5 The compilation of the user programs

The workstation in a cluster operates normally under the same operating system and on a compatible hardware platform. The same hardware and software platform makes it possible to compile the user program by the algorithm server. The source code is never transferred into the cluster in this case, only the binary code. If there are heterogen softwares or hardware platforms, than it is obviously not longer possible to transfer binaries ready to run, because a binary compiled e.g. for an Intel processor will not run on a Motorola processor. In this case the source code is transferred to the algorithm client. The algorithm client has to compile the source code at its own. This just-in-time-compiler tests the implementation date of the sourcecode and looks up the name and date in his precompiled application folder. If a precompiled version of the application exists in the application folder and its implementation date does not coincide with the date of the sourcecode, then the binary version is directly executed, but if there exists not a binary version or if this version is out of date, the source code is compiled in a transparent manner. Errors in the source are reported to the algorithm server. The algorithm server collects these errors and separates them into platform or version failure or semantical or syntactical errors. Identical errors are only reported once to the client.

A homogenious software and hardware platform is to prefere but not required.

1.6 Run control on the user program

The user programs on the workstations in the cluster depends on each other. The synchronization is one of the biggest problems of a cluster. Deathlocks are always possible. Hardware or software may occur at any time. A cluster must de able to recover from all these situations. The user should be able to stop a runaway process. Debugging possibilities are needed to fix bugs.

1.7 Heartbeat and negative acknowledge

The general synchronisation mechanism is based on a passive heartbeat synchronisation. The algorithm server broadcasts in a regular timeframe, a UDP-packet with a timestamp. The workstations are responsible to evaluate this timestamp and to compare it to their internal clock. If they miss several heartbeat packages they try to inform the server of this problem by a negative acknowledge packet, and if this fails they try to reconnect. However, the server may send a control block together with these synchronization packets. This control block may be a request to a client to report its status, to stop the execution, to pause the execution, or to continue the calculation. If a client does not answer such a request, the server assumes, that the client is no longer a member of the cluster and reconfigures the scheduling of the algorithm and deletes the corresponding workstation-ID from its internal tables. Pending algorithms from this workstation are send to other menbers of the clusters.

1.8 User Interface

The interface server interacts with the interface clients over the internet. The Apache web server sends a Java application to the clients web browser. This application extends the clients web browser capabilities to the support of a cluster user interface. This application together with the HTML forms give a user interface more sophisticated that these of other cluster packages like PVM or MPI.

Most workstations have already a web browser like the Internet Explorer or the Netscape Communicator installed. Together with the Java technologies RMI, Servlet, JSP and Webstart the remote developing and debugging is realized. The security of this application is trusted by a digital signature and tested every time, the application is downloaded. The sandbox security model of Java is extended where it was necessary. The user still has the full control over the security on his computer. Firewalls are tunneled over http. This assure the access of interface clients even from very restricted locations.

2. CONCLUSION AND PREVIEW

It is possible to create prediction systems based on neuronal networks for agricultural products, but these neuronal networks should be retrained with samples from the tested charge to give a good prediction probability quality. A cluster of workstations reduce the time of the retraining to an adequate frame, because the calculation of neuronal networks may be done on separate workstations at the same time. An embedded programming language developed here makes it possible to build programs running in the cluster without the knowledge of synchronization, or communication, or other in depth knowledge of cluster programming libraries. The building of a cluster has become easy and the resulting performance achieves the power of a mainframe to a part of the costs.

The cluster model may be used with similar problems on different agricultural products. The software will be extended in the future to integrate the power of external workstations and it should be possible to create a complete web site with a prediction system. The integration of a transcription server should improve the user interface.

On a sample of data as impact force index, temperature and others, some variants of neural networks are trained and their performance are evaluated. The evaluation factor is found to be about 60% to 70% for the prediction capability of the neural network to evaluate that the impact factor is the most relevant influencing factor on the development of black spots on potatoes for any certain variety.

REFERENCES

Anonymous, ANTLR, Scanner and Parser for Java http://www.beowulf.org

Brause Rüdiger (1995). *Neuronale Netze*, B. G. Teubner, Stuttgart.

Haykin Simon (1994). *Neuronal Networks - a comprehensive foundation*, Prentice Hall.

Richter S., Gottschalk K. (2000). Schwarzfleckigkeit von Kartoffeln, *Landtechnik* **3/2000**

Wormanns, G., Jacobs, A. (1997). Methodik und erste Ergebnisse zur Simulation der mechanischen Belastung von Kartoffeln (Methodology and first results of the simulation of mechanical stress on potatoes). In: *Landtechnik: Tagung,* VDI-Berichte 1356, Braunschweig.

NEURAL NETWORK MODELLING FOR PREDICTING RAINFALL LEVEL

Ku Ruhana Ku-Mahamud, Yuhanis Yusof, Teoh Boon Wei

School of Information Technology, Universiti Utara Malaysia
06010 Sintok, Kedah
Malaysia

Abstract: The back propagation neural network model to predict the rainfall level for two states has been developed using C and Visual Basic programming languages. Rainfall levels from 56 rainfall stations for 29 years (1970-1998) has been used and the results obtained through the use of neural network were compared to the one obtained from regression. The study has succeeded in initiating an application to predict the rainfall level for a 10 year period. *Copyright © 2001 IFAC*

Keyword : Artificial Intelligence, Neural Network, Back Propagation, Prediction, Training.

1 INTRODUCTION

The irrigation management system of Kedah and Perlis, which is maintained by Muda Agricultural Development Authority (MADA) is the biggest irrigation management system in Malaysia. There are 73 rainfall collection centers or stations monitored by MADA throughout Kedah/Perlis, through the Water Management and Control Scheme (WMCS). The WMCS is used to control the flow of water as well as to establish a system of monitoring, controlling, managing and evaluation of the water resources utilization in Kedah and Perlis.

Each station collects an represents the water level of the rainfall for an area of 16 km^2. The water sources are rainfall, dams, rivers and 12 recycle stations scattered around the region. There are three dams available in Kedah, namely Muda, Ahning and Pedu. Out of the four water sources mentioned above, rainfall constitutes more than 50% of the total water source to the states. As in MADA's context, water management means the supply, distribution and application of the right amount of water at the right time, to the right place so that the paddy plants would thrive and produce good yield (Teoh and Chua 1989).

Rain prediction has always been a critical issue in an equatorial country like Malaysia, since the rain affects a lot of its peoples' activities. If the rainfall level can be predicted accurately, proper planning can be carried out to facilitate the running of rain-associated activities. For instance, farmers can plan their paddy plantation according to the quantity of rain in certain periods in a particular area. Despite the significant role of the rainfall data in the WMCS, no attempt at present has been made to forecast the rainfall level so as to help the agriculturist to plan ahead the paddy plantation activities. This is because prediction of rain is difficult and inaccurate due to the fuzzy and complex underlying relationship between the various determinant factors that effect rain fall.

Five methods have been used to forecast the rainfall level as reported by the University of Illinois (1999). The first of these methods is the persistence method. The assumption made while using this method is that the conditions at the time of the forecast will not change. This method works well if the weather patterns change very little and the features on weather maps move very slowly. The persistence method can be used in predicting long range weather conditions or making climate forecasts.

The second method determines the speed and direction of movement of the pressure centers, areas of the cloud and rainfall level before making the forecast. By using this information, the forecaster attempts to predict where those features will appear at some point in the future. This method works well when the weather systems continue to move at the same speed and in the same direction for a long period of time.

Climatology is another method that has been used by the University of Illinois and it involves averaging weather statistics accumulated over many years to make the forecast. This method works well if the rainfall pattern is similar to the chosen time of the year.

The fourth method is the analog where it examines the day's forecast scenario and attempts to retrieve another day in the past where the weather scenario looked similar. This method is difficult to use since it is hard to find a perfect analog as weather features seldom repeat. Even small differences between the predicted time and the analog can lead to a whole range of differences.

The fifth method is numerical weather prediction. This method makes use of high power computers to make a forecast. Computer programs will be used to simulate the atmosphere. Variables such as rainfall, temperature, pressure and wind conditions can be computed. The forecaster then produces the weather forecast for a particular day based on the atmospheric variables computed.

Area time integral has been used by Brandes (1998) to estimate rainfall level. To adopt this method, the probability of the density function of the rain level for a particular region must be available. The average rain levels are then correlated with the fraction area of rain levels above a specific threshold. The proportionality factor of the correlation can be determined with disdrometers or radar reflectivity. This method is ideal for predicting the mean of the rainfall level for a specific time interval.

The back propagation neural network (NN) approach has been used to predict the rainfall level in this study. Findings from this study will provide better rainfall prediction and thus will help the authority to effectively control the release of water and will also enhance the usefulness of the Kedah WMCS.

2 METHODOLOGY

There are five phases in building the back propagation NN application to predict the rainfall level. The phases are network design, data preprocessing, network training, validating and testing, development of the user interface and result validation.

2.1 *Network Design*

In view of the nature of the problem, back propagation network and forward-mapping counter propagation network were considered, since the architecture satisfied the mapping requirement of the application. In accordance to the data obtained for the 56 stations, it was decided to use three nodes for the input layer. These three nodes represent the stations where the data was collected, the year and the month. The number of nodes in the output layer is one and that represents the amount of rain fallen. A single hidden layer has been used for the network. Since the application contains only three nodes in the input layer, only two nodes were built in the hidden layer.

2.2 *Data Preprocessing*

Data obtained was represented, scaled or normalized, randomized and finally segmented before being used. The stations have been assigned a code from 1 to 56. As for the months, January has been assigned the code '1', February with '2' and it goes on until December, the assigned code is '12'. After the data was represented, it was then scaled down to values in the range between 0 and 1. NNs are sensitive to absolute magnitudes. Each of the value for the rainfall station was then divided by 56 and each of the value for the month was divided by 12. The values for the year and the rainfall level were also scaled down to values in the range between 0.1 and 0.9. New values for the year and rainfall level were obtained by applying a formula as given by equation 1 (Skapura, 1996).

Equation 1

$$x' = \frac{(x - x_{min}) * 0.8}{(x_{max} - x_{min})} + 0.1$$

Data was randomized and segmented into different groups. This was carried out in the following manner. There were 17952 sets of input pattern for training purposes. These sets were divided into four groups namely A, B, C and D whereby each group contained 4488 sets of data.

2.3 *Network Training, Validating and Testing*

For each group, 90% of the data was used for training and the remaining 10% was used for validating. If group A was used for training and validating of the model, then group B, C and D were used for testing the model produced by data from group A. The prediction accuracies of the model using groups B, C and D's data were recorded. The average prediction accuracy of the group was then calculated. The same training, validating and testing of the model for group A was then carried out on the other groups of data. The group of data that produced the highest average for correct prediction was then selected to be the best model.

The prediction program was developed using C programming language. Connection weights were initialized before the training started. A common procedure in choosing the weights between the range –0.5 and 0.5 (Fausett, 1994) was adopted and the formula that was used to generate weights is as in equation 2.

Equation 2

$$\text{weight} = \frac{\text{rand}}{\text{RAND\_MAX}} - 0.5$$

The *rand* is a C built in function that returns a uniform pseudo-random number in the range of zero through RAND_MAX. RAND_MAX is a constant value and its minimum value is 32,767.

2.4 *Development of the User Interface*

The user interface has been developed using Visual Basic to enable the user to predict the rainfall level for the years 1999-2009. The interface enables the user to input the station name, the year and the month and the program will output the predicted rainfall level. Results of the prediction will be displayed in both text and graphic form.

2.5 *Result Validation*

The results obtained using the NN approach has been compared to the one produced by regression analysis. The regression analysis was done through the use of least square method.

3 FINDINGS

Table 1 depicted the results from the training and testing phases of the network .

Table 1 : Testing Result for Group A

Training & Validating Group	A	
Testing Group	B, C, D	
MSE	0.009333	
Generalization Error	0.008108	
Connection Weights		
Input 1 – Hidden 1	-1.602029	
Input 1 – Hidden 2	0.306776	
Input 2 – Hidden 1	-2.645369	
Input 2 – Hidden 2	0.191145	
Input 3 – Hidden 1	3.06955	
Input 3 – Hidden 2	-6.44061	
Hidden 1 – Output	-0.996928	
Hidden 2 - Output	-4.467729	
Epoch	569	
Testing Result	Set	%
	B	72.75
	C	72.50
	D	72.30
Average Testing Result	72.44 %	

The best model obtained was the one that has been trained and validated by data from group A with 72.44 % correct prediction.

Figure 1 shows the percentage of correct prediction during the testing phase. It is interesting to note that the percentage of correct prediction were almost the same with four different training sets of data. This indicates that all the data was randomly distributed.

Prediction Accuracy

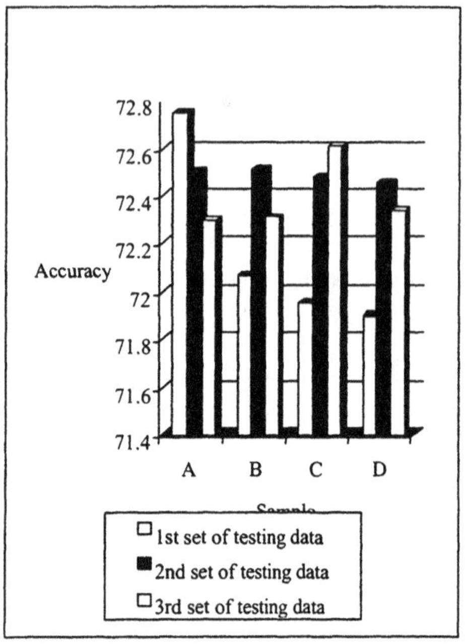

Figure 1 : Group Prediction Accuracy

The input month had significant effects on the adjustments of the weights as compared to other inputs (refer Table 1). The connection weights of input node 3 to hidden node 1 and from input node 3 to hidden node 2 were 3.069550 and -6.440061 respectively. In other words, the variation of the rainfall level could be traced monthly and not by the year or location.

The regression analysis has generated the following prediction model.

Equation 3

$$y = 0.0218187\ x_1 + 0.0674073\ x_2 + 0.2176557\ x_3 + 0.1348864$$

where y is the rainfall level, x_1 refers to the station or location, x_2 refers to the year and x_3 is the month. The above model has been generated from the data of group A. The model was tested against other groups of data and the percentage of correct prediction was 69 %. The NN model has better prediction result compared to the regression analysis approach. Figure 2 depicted the architecture of the best model.

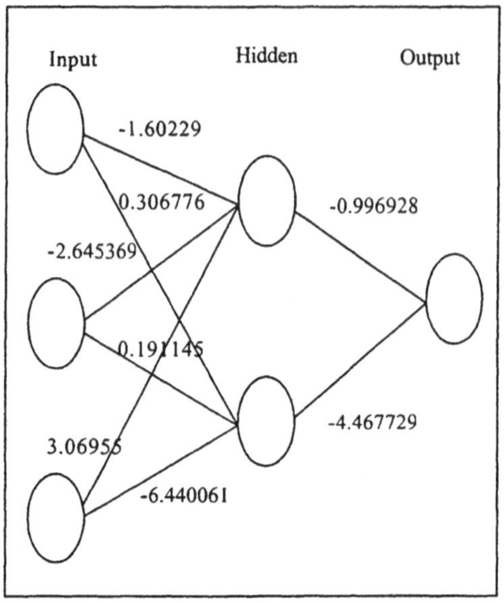

Figure 2 : Best Network Model Architecture

4 CONCLUSION

This study has succeeded in initiating an application to predict the rainfall level in the area under the MADA irrigation management system. The back propagation NN has demonstrated its ability in predicting the rainfall level. Results obtained through the use of the NN approach is superior compared to the one obtained from regression analysis. The program produced can help in creating a sound irrigation system in managing and enhancing the usefulness of the Water Management and Control Scheme of Kedah.

REFERENCES

Brandes, E.A. (1998) A Review of Research and Development Activities Related to WSR-88D Algorithms. [www document], URL, http://www.osfnoaa.gov/app/sta/algorithm98.htm #2.6PrecipitationAnalysis.

Fausett, L. (1994) *Fundamentals of Neural Networks: Architecture, Algorithms and Applications, New Jersey: Prentice Hall.*

Skapura, D.M. (1996) *Building Neural Networks,* New York: Addison-Wesley Publishing Company.

Teoh, W.C. & Chua, T.S. (1989) *Irrigation Management Practices in MADA*, Jabatan Parit dan Taliair Malaysia.

University of Illinois (1999) The Weather World 2010 Project. [www document], URL, http://ww2010.atmos.uius.edu/(GH)/guides/mtr/fcts/mth/oth.rxml.

NEURAL NETWORK MODELLING OF SOLAR COLLECTORS

István Farkas[1], Piroska Géczy-Víg[1], Mihály Tóth[2]

[1]Department of Physics and Process Control, Szent István University
Páter K. u. 1., H-2103 Gödöllö, Hungary
E-mail: ifarkas@fft.gau.hu

[2]Process Control Research Group, Hungarian Academy of Sciences SZIUG
Páter K. u. 1., Gödöllő, H-2103 Hungary

Abstract: In this study different modelling approaches of flat plate solar collectors are introduced and analysed. Among the physically based models the heat network technique and the Hottel-Vhillier models are discussed. The parameters of the latter one are identified for three different type of flat plate solar collectors. The identification showed a good coincidence with the measured values. Finally, modelling trials with a neural network (NN) technique were carried out. A sensitivity study were performed with the parameters of the neural network. Beside the possible NN structures, the size of training data set, the number of hidden neurons, the type of training algorithm were analyzed in order to get the most appropriate model. The same structure of NN were trained successfully for the studied three different type of flat plate solar collectors. *Copyright © 2001 IFAC*

Keywords: Solar collector, modelling, heat network, identification, neural network

1. INTRODUCTION

The flat plate solar collectors are commonly used in the different areas of farm operation, that is for making hot water, for drying of agricultural materials, etc. During the design and controlled operation of such equipment it is often required to compare the performance of several types of solar collectors taking into account the thermal efficiency and the economical aspects, too. Therefore, to develop an appropriate model and carry out its parameter identification is a rather important task. At the same time the control issues of the collectors are to be considered as well.

In the literature several aspects of such a modeling technique are referred. Pierson and Padet (1990) worked out a method for determination of time constant of solar collectors to be used in their transient modeling.

Hashis and El-Refaie (1983) applied Pad; approximation to obtain a reduced order model from the partial differential equation system representing the dynamic behavior of a flat plate solar collector.

Prapas et al, (1988) modeled the transient behavior of solar collectors with introducing a response function. The method is especially applicable when the prediction is required at a small time intervals.

In the recent study different modelling approaches of the thermal behaviors of solar collectors are carried out. First, the neat network modelling technique is introduced which could serve as a basis for another physically based model, as for instance, the Hottel-Vhillier one. This latter one is simply enough to use even for control purposes of the collector operation. The identification of the Hottel-Vhillier model for different flat plate collectors can be performed via measurements. And finally, the neural network model are to be considered to use for modelling purposes. The training of NN can be done via measured or simulated data.

2. HEAT NETWORK MODELLING

To calculate the simultaneous heat and mass transport in a solar collector the discretized models can be applied successfully. It means, the collector is divided into typical subparts, in which a uniform temperature distribution is assumed. In Fig. 1. the cross section of a flat plate solar collector is drawn giving a possible discretization.

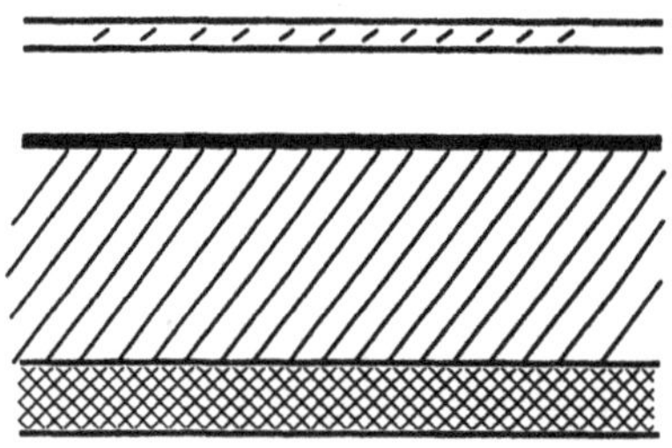

Fig. 1. Cross section of a flat plate solar collector

The accuracy of these models depends on the degree of discretization along the length, as because the most important temperature changes are only along the length of the collector not taking into consideration the heat flow along the width, i.e. the heat loss at the side of the collector.

To describe the corresponding heat and mass balance equations with the relating initial and boundary conditions the heat flow network (HFN) method can use successfully for such modelling as it was discussed in Farkas (1998) and Imre et al, (1985). The HFN includes the heat resources and the heat capacitance of the discretized part and also the heat transfer resistance between the individual elements. In case of solar collectors the main energy gain comes from the solar radiation, and the thermal connection to the ambient is taken into consideration. The advantage of the HFN is that it can be generally formulated and so the same simulation model can be used for different structures and for different discretizations achieving the required accuracy. The HFN model of the collector shown before is illustrated in Fig. 2.

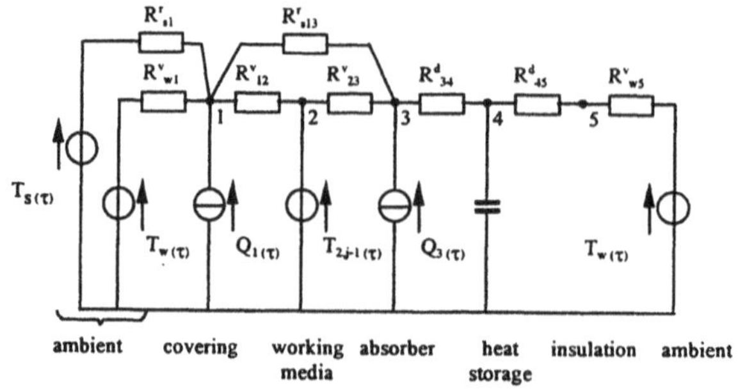

Fig. 2. Heat flow network of a flat plate solar collector

In this model the heat capacitance of the subparts are neglected except the heat storage. The HFN model is valid for the i-th part of the collector along its length. Therefore, at the node 2 is expected to insert a temperature generator with a value yielding from the previous (i-1)-th section. On the basis of model assumptions, however, the calculations can be done with one network, and their connections could take into consideration by means of the actual heat capacity flow.

The HFN model, at the same time, could serve a basis to optimal operation of a solar system and to determination the parameters of a reduced order dynamic model sufficient for control purposes.

Generally, the control strategies applied for regulation of flow of working media in a flat plate collector use a simple dynamic model. The main characteristics of such a dynamic model include the effect of overall collector capacitance and the heat transfer coefficients between the absorber and the environment and between the working media and the environment, as well. That can be assumed as a specific case of a HFN model with two nodes.

3. HOTTEL-VHILLIER MODEL

The application of the solar collector determines what model is used for its study. In the many of practical cases, especially in the agricultural ones even for control, it is enough to apply a low order dynamic model to be identified. Keeping in our mind these statements, it is often decided to take into consideration the Hottel-Vhillier model, to calculate the temperature distribution i.e.:

$$T(x,\tau) = T_w(\tau) + I(\tau)/k_{aw} + [T_{in}(\tau) - T_w(\tau) - I(\tau)/k_{aw}] \exp[-k_{mw}\, w\, x / (c_m \dot{m})], \qquad (1)$$

where:

T - collector outlet temperature (°C),
X - length coordinate (m),
τ - time (s),
I - solar radiation (W/m²),

T_w - ambient temperature (°C),
T_{in} - inlet temperature (°C),
w - width of the collector (m),
c_m - specific heat of medium (J/kgK),
$\dot{m}$ - mass flow rate of medium (kg/s).

This first order model is quite simple and it serves only the temperature distribution of the flowing media along the length of collector. At the same time, it is rather difficult precisely to determine the heat transfer parameters, k_{aw} and k_{mw}, on physical basis. But, to identify these parameters via measurements or via results obtained from the detailed HFN model seems to be a promising solution.

The identification problem of such leads to minimize the calculated (c) and measured (m) temperature difference over the operating time (τ) interval:

$$J[k_{aw}, k_{mw}] = \int_{\tau_0}^{\tau_e} |T_c(\tau) - T_m(\tau)| \, d\tau \qquad (2)$$

Furthermore, the identification results based on the measurements in Imre et al, (1981-84) are shown for three different collectors using a multi-parameter optimization.

3.1. Air collector

The collector was horizontally positioned and covered by a hard foil sheet (lexane). The main data were as follows: length 1.91 m, width 0.95 m and the air flow rate 0.053 kg/s. The measurement has been carried out on a cloudy day in June which can be observed from the ambient and inlet temperatures. Using the Eqs (1), (2) the heat transfer parameters were identified as:

k_{mw} = 56.519 W/m^2 K and
k_{aw} = 37.893 W/m^2 K.

The final results are shown in Fig. 3. where the outlet temperature calculation was done by means of the identified parameters.

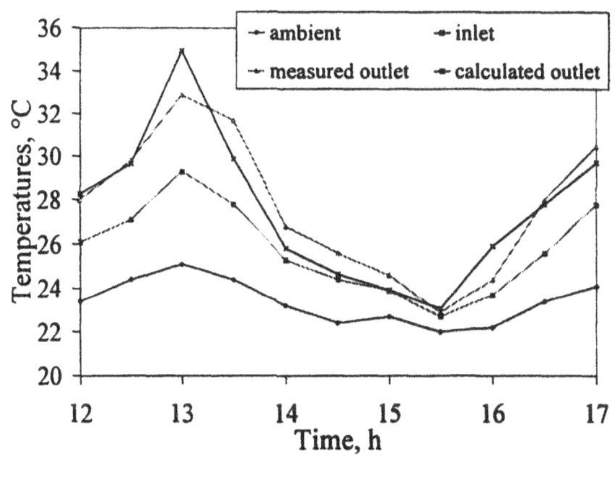

Fig. 3. Identification of an air collector

The maximum difference between the measured and calculated outlet temperatures was about 2 °C which occurs after the sharp changing in cloudiness. Otherwise, the estimation reasonably fits well.

3.2. Water collector

The water collector was also horizontally positioned and covered by a hardfoil sheet. The main data were as follows: length 1.91 m, width 0.95 m and the air mass flow rate 0.013 kg/s. The measurement was performed been carried out on a clear day in September. Using again the Eqs (1), (2) the heat transfer parameters were identified as:

k_{mw} = 66.205 W/m^2 K and
k_{aw} = 20.576 W/m^2 K.

The final results are shown in Fig. 4.

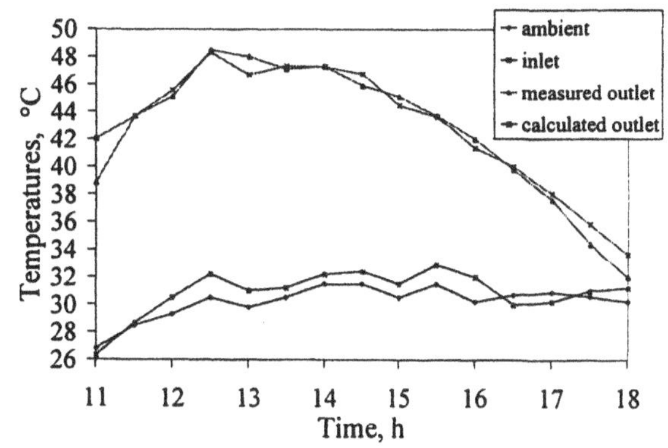

Fig. 4. Identification of a water collector

The maximum difference between the measured and calculated outlet temperatures was about 1.4 °C. The estimation shows a very good coincidence in the entire time interval.

3.3. Heat storage collector

A latent heat storage matrix of $CaCl_2 \times 6 H_2O$ was used to build up the absorber of the collector. Its critical phase change temperature is 29 °C and a band of ΔT=±1 °C was used around that value for the estimation of the phase change process. Otherwise, the geometry and the position of the collector were the same as in the previous experiments. The air flow rate was 0.057 kg/s. The measurement has been carried out on a clear day in May with some cloudiness which can be observed from the ambient and inlet temperatures. Using again the Eqs (1), (2) the heat transfer parameters were identified as:

k_{mw} = 55.042 W/m^2 K and
k_{aw} = 42.820 W/m^2 K.

The final results are presented in Fig. 5.

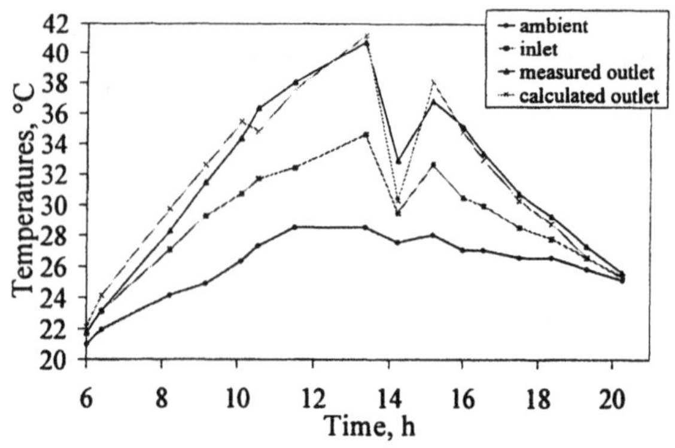

Fig. 5. Identification of a latent heat storage collector

The maximum difference between the measured and calculated outlet temperatures is about 2.5 °C, which occurs at the sharp cloudiness changing only. Otherwise, the estimated values fit quite well for the whole time interval, even thought, no any corrections were used due to the latent heat storage effect.

4. NEURAL NETWORK MODEL

The next step of the recent study was to carry on with some modelling trials concerning to the thermal behavior of the solar collector using artificial neural network (NN). The simplified structure of the neural network to be used for modelling of a solar collector is shown in Fig. 6.

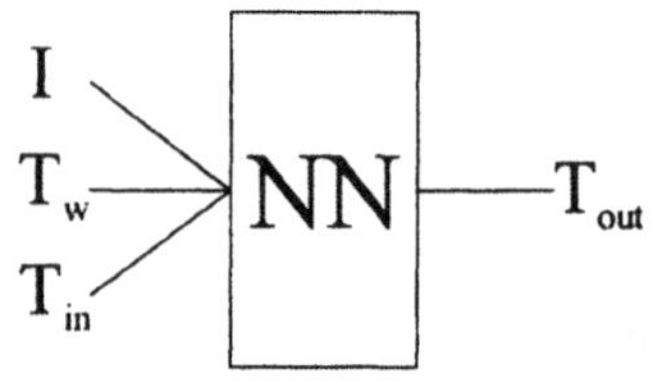

Fig. 6. Simplified NN structure used for modelling of solar collectors

The first layer of the NN includes the inlet parameters of the collector as the solar radiation, the ambient temperature and the temperature of the inlet working medium. The last layer of the NN should contain one neuron because there is one output parameter. The problem could not be solved, however, with one neuron so the next most simple possibility to be examined is that the neurons should be positioned in two layers, but using one neuron in the second layer with linear step function (purelin).

The size of the ideal training data set, the number of hidden neurons, the training algorithm and the step function should be determined for the first, hidden layer.

The detailed examination was made for the heat storage collector with the Matlab+Simulink software along with the Neural Network Toolbox. Concerning to the training the different variations of the Back propagation algorithm were tested. The success of training was characterized by the average of mean square deviation between the measured and the calculated values of the outlet temperature (T_{out}) of the solar collector and the values generated by neural network. The minimization of this mean square deviation was the main purpose.

4.1. The size of the training data set

The data from measurements (16 data pairs) turned out to be insufficient. The outcoming results were very different repeating the training process with the same number of neurons, training algorithm and step function.

Intermediate values were generated for the measured input data by linear interpolation from which the T_{out} values were calculated by the Hottel-Vhillier model. The examined number of elements in the training set were: 16, 88, 117, 218, 347, 1039 and 5171. In the case of 218 element set the error was within 0,22 °C. The training process got slower as the size of the set got larger and the approximation did not get significantly better. The deviation lessened minimally after the first 50 training cycle.

4.2. The number of hidden neurons

The Table 1 below shows the average of mean deviation from more training results by the number of neurons.

Table 1 Influence of neuron numbers

Number of neurons	Mean square error
3	0,07
4	0,05
5	0,04
6	0,04
7	0,02
8	0,025
9	0,03
10	0,03
15	0,03
20	0,02
40	0,02

The deviation and the time of the process are both acceptable when the number of neurons is around 7-8. Finally, it has been decide to use 7 neurons for further examination.

4.3. Step function

There were several options what kind of step function was to be used. From a choice of hard-limit (hardlim), tan-sigmoid (tansig), log-sigmoid (logsig) and linear (purelin) functions, finally, the tansig function proved to be the best for our case, i.e. for solar collectors.

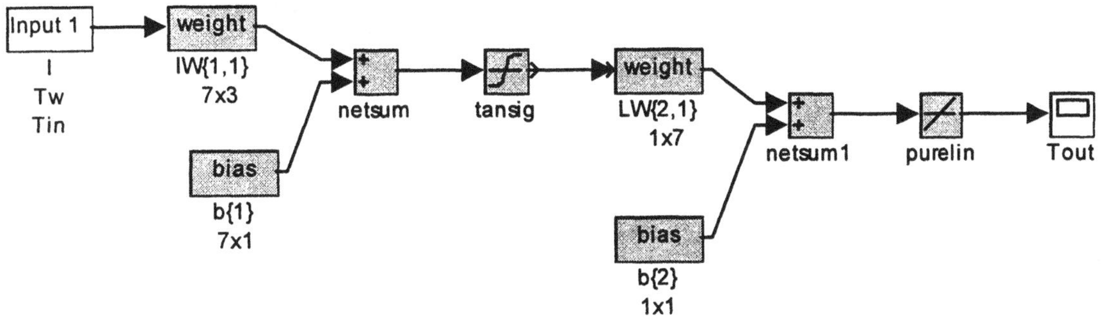

Fig. 7. Detailed NN structure used for modelling of solar collectors

4.4. Training algorithm

The software has 11 variations of the back propagation algorithm, from which only the Levenberg-Marqardt algorithm produced acceptable approximation (trainlm).

Considering all the above mentioned facts the finally selected structure of the optimal neural network can be seen in Fig. 7.

5. TRAINING RESULTS OF NN

As a first resort, training of the ideally selected neural network structure was carried out for three different kind of solar collectors introduced in the chapter No.3. During the training all the parameters of the NN were kept identically. The collector outlet temperatures (T_{out}) were generated with the Hottel-Vhillier model.

The training results of the neural network for the heat storage air collector are shown in Fig. 8.

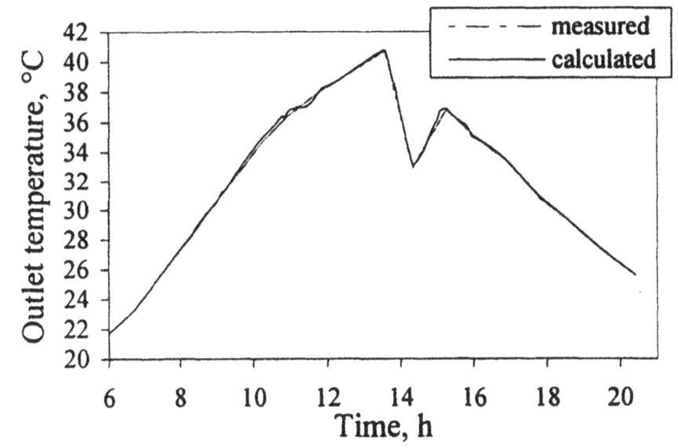

Fig. 8. NN training results for latent heat storage collector

In the figure a very nice coincidence between the training and the NN produced date can be observed. The NN structure proved to be optimal can be used for modelling the other two types of collectors (air collector and water collector), but of course the value of weight matrix and bias vector is different in those cases.

Just for the case, a trial was made to train the neural network only with the limited number of measured data pairs. In spite of the low number of data the training was rather successful. The result of such training for the air, for the water and for the latent heat storage collector are shown respectively in Figs 9-11.

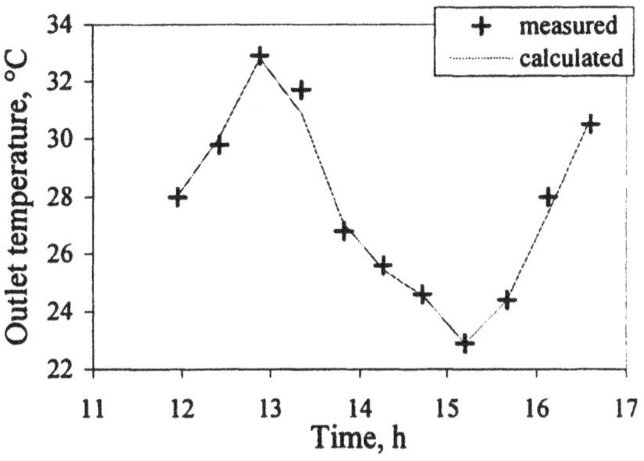

Fig. 9. NN training via measurement for air collector

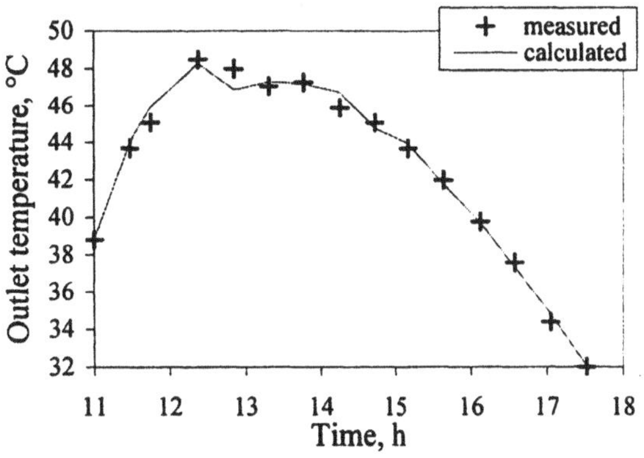

Fig. 10. NN training via measurement for water collector

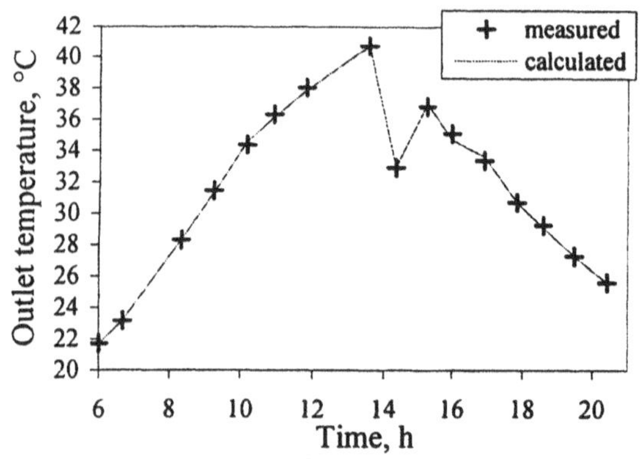

Fig. 11. NN training via measurement for latent heat storage collector

6. CONCLUSION

In this paper different modelling approaches were studied for the determination of the thermal behavior of flat plate solar collectors.

It can be stated that the simplified physically based model, as the Hottel-Vhillier one can be successfully identified through measurements or using the more detailed heat network models (HFN).

Concerning to the applicability of the artificial neural networks for solar collector modelling, first, an ideally structure has to be developed. The NN is rather sensitive to the size of the training data set, the number of hidden neurons and the training algorithms, as well.

The ideally selected NN structure was successfully trained for three different kind of solar collectors (air, water and latent heat storage) using the measured and also the generated data provided by the previously identified Hottel-Vhillier model.

The validation of the trained model requires further studies.

Acknowledgements: This study was carried out with the support of projects TET D-17/1998, OTKA T-029300 and T-032510.

REFERENCES

Farkas,I.: Identification of flat plate solar collectors used in agriculture, 9th IFAC-IFORS Symposium, Budapest, Hungary, July 8-12, 1991, Preprints, Vol. 1, /ed. by Cs. Bányász and L. Keviczky/, p. 475-479.

Farkas, I. /ed/ (1998), Modelling, control and optimization. Greenhouse, drying and farm energy system, Gödöllő University of Agricultural Sciences, Textbook, Gödöllő, Hungary.

Hashish, M.A. and El-Refaie, M.F. (1983), Reduced order dynamic model of the flat-plate solar collector, *Applied Mathematics Modelling*, **Vol. 7**, p. 2-10.

Imre, L. /topic leader/ (1981-1984), Solar agricultural crop dryer, Research Report, Technical University Budapest.

Pierson, P. and Padet, J. (1990), Time constant of solar collectors, *Solar Energy*, **Vol. 44, No. 2**, p. 109-114.

Prapas, D.E., Norton, B., Milonidis, E. and Probert, S.D. (1988), Response function for solar-energy collectors, *Solar Energy*, **Vol. 40, No. 4**, p. 371-383.

COLOR AND SHAPE ANALYSIS BY NEURAL NETWORK

Anett Szepes

Department of Physics & Control, UHFI
H-1118 Budapest, Somlói út 14-16, E-mail: anett@elfiz2.kee.hu

Abstract: Nowadays the traditional methods in the area of classification are the statistical methods like cluster analysis or discriminant analysis. The use of neural network is unusual, but it can give better results in classification than statistical methods. In our research we analysed three classification methods: cluster analysis, discriminant analysis and neural network when two kind of parameters were given to us from the analysed agricultural produces. These parameters were the color and the shape, which were obtain from the computer vision system. *Copyright© 2001 IFAC*

Keywords: Neural network, Discriminant analysis, classification, color, shape

1. INTRODUCTION

Classification analysis is based on classical statistical method generally. Majority of the statistical methods are in practice like cluster analysis or discriminant analysis. The cluster analysis within the K-means cluster analysis is the one of the simpliest way to classify a set. This is a non-parametric method to directly classify the set. An other group of statistical processes is the parametric method, where a classification criteria helps the classification. The neural network is not a traditional method to classify the analysed set. The work of neural network is unusual, but it needs less time to classify than other statistical methods. This method can be automated to the industry. In this process the results of image analysis (color and shape properties) can be used to give an objective results.

Our first objective was to analyse the goodness of the use of neural network in the cases of agricultural processes.

The second objective was to develop a method for color and shape characterisation based on neural network.

2. MATERIALS AND METHODS

2.1. Materials

The produces tested were as follows:

Five pear varieties were: „Beurre Durandeau" - 14 pieces, „Vilmos" – 14 pieces, „Hoshui" – 16 pieces, „Peckham Triumph" – 12 pieces- and „Clapp kedveltje" – 16 pieces.

Three cereal varieties: wheat, lightseed and hemp - 50-50 seeds.
Two tomato varieties: „Heinz" – 145 pieces, "Ispan" – 46 pieces.

2.2. Measuring system

A lighting system, a video camera, a frame grabber and a PC were in a measuring system developed earlier (Felfödi, *et. al.*, 1994).

To take the photographs four ordinary lamps were used. The surface of tested produces was chosen according to the average color. In the case of cereals, the photo was taken from the shade of seed. That is to say the seed was between the camera and the white paper sheet, which was lighted by lamp.

2.3. Shape and color characterisation

Fast Fourier transformation and PQS (Polar Qualification System) method (Martinovich and Felföldi, 1996) were used to characterise the shape features of pears and cereals. The results of FFT were the first four Fourier coefficients (Székely, 1994), characterising the elliptic, triangular, quadratic symmetry of the shape. The PQS method is based on determining of centre of examined object. The color features were examined by a computer software, which can give the average value of the R, G, B parameters of the tested surface (Felföldi *et. al.*, 1994). In the case of tomato the value of R, G, B parameters were transformed to L, a* and b* parameters in CIELab color system.

2.4. Statistical methods

Three methods were used to evaluate the results of image processing.

Cluster analysis: The K-means cluster analysis was used in the case of tomato cultivars. This was the first step for dividing the group without any classification criteria.

Discriminant analysis: At this method 2 parts were created from the sample. The first part is the training sample. This sample has to be classified by experts before analysing with discriminant analysis. Some classifying functions were created which based on training sample. The parameters of the functions were the Fisher's coefficients. The other part of the sample, which is the test sample, was classified on the basis of classifying functions.

Artificial neural network: Actually this method isn't a multivariate statistical method, but it was used as classifying method. The artificial neural network is similar to human neural network and its elements resemble the neurone. This simple model works with a threshold function based on weighted average of input data.(Simpson, 1990) The Figure 1. shows the work of the network's element.

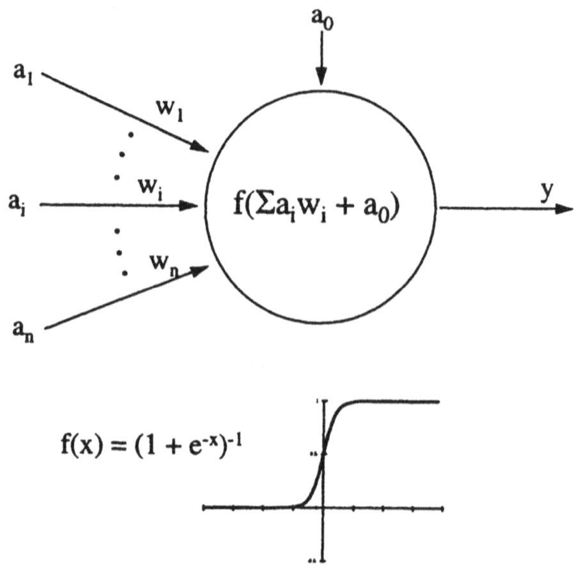

Fig. 1. The work of element of the Neural Network

The elements of a network are in one field. That field receives the input data called input layer, that field gives the result called output layer and that field which is between the input and output layer is called hidden layer.

Figure 2. shows the structure of a three layers Neural Network. The input data were the examined parameters and the output data were the classes of fruit and vegetable.

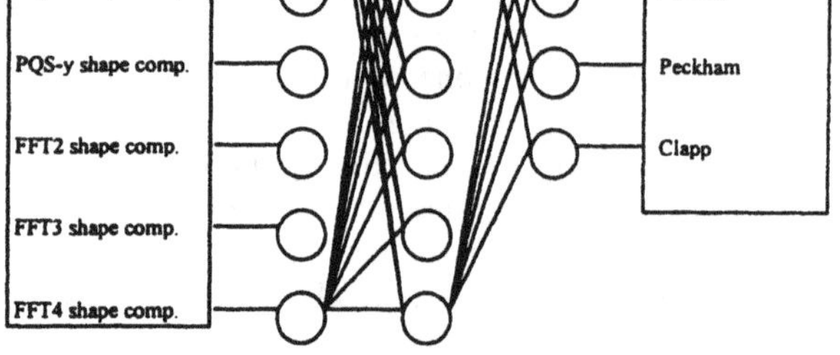

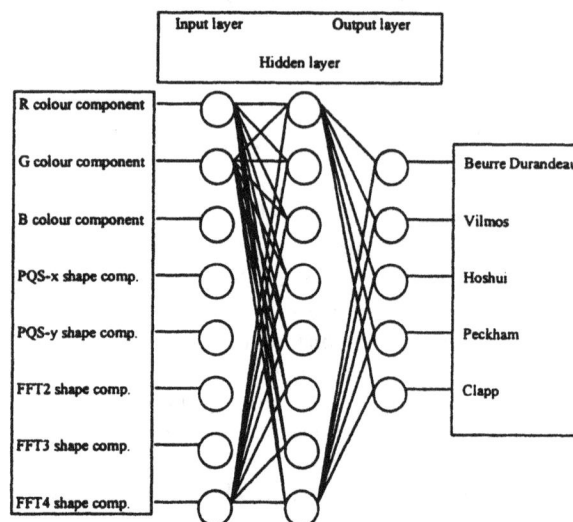

Figure 2. The structure of network in case of pears

Multilayer Backpropagation Neural Network was used which is developed at Department Physics & Control of Szent István University (Firtha, 1996).

The linear combination of output state of previous elements is the input data of the next layer elements. The backpropagation network gives a difference between the output data and the target. This difference value will be an input data too. The output data is between 0 and 1 according to threshold functions. (If other kind of threshold functions are used, the output value can be between –1 and 1.) The best fit of calculated value to the target group means the 1 value.

At the application of neural network we must use 2 sample groups. One of them is the learning sample. The neural network is created during the learning. This process is based on iterations. One iteration means the combination of the input data. The running time is approximately 32s/5000 iterations with pear cultivars. The other part is the test sample which is unknown for the ready network. The goodness of network classification was measured in application of test sample.

3. RESULTS

The shape parameters were analysed with the pear and cereal cultivars. The shape parameters were the PQS-x, -y components and the FFT2, FFT3 and FFT4 components.
The classification of the learning sample by the Discriminant Analysis and Neural network were good, but in the case of test sample it was not sufficient. Table 1 shows the result of classification of pear test sample by Neural Network.

Table 1. Classification of test sample of pears by Neural Network

	Predicted				
Original	Beurre Durande	Vilmos	Hoshui	Peckham	Clapp kedveltje
Beurre Durandeau	5	0	0	1	1
Vilmos	0	4	0	0	3
Hoshui	0	0	8	0	0
Peckham	2	1	0	1	2
Clapp kedveltje	1	0	0	2	4

The rows are the target groups; the columns show the results of classification. The results of the classifying of Discriminant Analysis was similar to the result of Neural network. Only the Hoshui cultivar classifying rate was 100%. But in this case this cultivar's shape can give difference from other four cultivars without any difficulties.

The results obtained with cereals by Discriminant Analysis are shown in Table 2.

Table 2. Classifying of test sample of cereals by Discriminant Analysis

	Predicted		
Originals	Wheat	Lightseed	Hemp
Wheat	23 95,84%	1 4,16%	0 0,00%
Lightseed	0 0,00%	25 100,00%	0 0,00%
Hemp	0 0,00%	1 4,00%	24 96,00%

Two seeds were misclassified: one from wheat to lightseed and the other from hemp to lightseed. The Neural network gave the results shown in Table 3.

Table 3. Classification of cereal test sample by Neural Network

	Predicted		
Original	Wheat	Lightseed	Hemp
Wheat	24 100,00%	0 0,00%	0 0,00%
Lightseed	0 0,00%	25 100,00%	0 0,00%
Hemp	0 0,00%	0 0,00%	25 100,00%

The result of classification was 100%. But in this method the selection of learning sample is very important. Because a not traditional seed-shape from the same cultivar can give an other kind of structure of Neural network, therefore the classification of test sample could give an other result, it depends on the aim of calssification.

In the case of pear samples the classification of shape parameters was only poor. Therefor the color parameters were added to the shape parameters. The result of classification are shown in Table 4.

Table 4. Classification of test pear samples by Discriminant Analysis based on color and shape parameters

	Predicted				
Original	Beurre Durande	Vilmos	Hoshui	Peckham	Clapp kedveltje
Beurre Durandeau	7	0	0	0	0
Vilmos	0	7	0	0	0
Hoshui	0	0	8	0	0
Peckham	0	1	0	5	0
Clapp kedveltje	0	0	0	0	8

One pear was misclassified from cultivar Peckham to cultivar Vilmos. The method of Neural network the classification was worse than in the method Discriminant Analysis. 2 pears were misclassified from cultivar Beurre Durandeau to cultivar Clapp kedveltje and 1 pear was misclassified from cultivar Peckham to ciltivar Vilmos.

With Neural Network two more pears were misclassified. The same Peckham pears were misclassified too. This special pear's shape and color resemble better to an ordinary Vilmos pear than to an ordinary Peckham Triumph pear. Because this pear's shape is more similar than cultivar Peckham, its color is more yellow than an ordinary Peckham Triumph which is green.

In the tomato sample 3 different colors were taken into account: green, pink and red. These different colors were in different groups.

In the case of tomato cultivars the first step was the using of K-means cluster analysis. The number of misclassified tomatoes was the biggest among the used methods. 6 misclassified tomatoes were in the tested sample from these tomatoes 3 were misclassified from green tomatoes to pink tomatoes. In the case of other two methods the classification by Neural Network was more successful than the classification by Discriminant Analysis.

The results of classification of "Heinz" tomato test samples by Discriminant Analysis are shown in Table 5.

Table 5. Classification of tomato test samples by Discriminant Analysis

	Predicted		
Original	Green	Pink	Red
Green	13	1	0
	92,86%	7,14%	0,00%
Pink	0	14	0
	0,00%	100,00%	0,00%
Red	0	2	40
	0,00%	4,76%	95,24%

With this method three tomatoes were misclassified, but in the case of classification by Neural Network, the classification gave a better result. None of tomatoes were misclassified that is shown in Table 6.

Table 6. Classifying of test sample of tomato in Neural Network

	Predicted		
Original	Green	Pink	Red
Green	14	0	0
	100,00%	0,00%	0,00%
Pink	0	14	0
	0,00%	100,00%	0,00%
Red	0	0	42
	0,00%	0,00%	100,00%

In the case of cultivar "Ispan" the results were similar than the cultivar "Heinz". The Discriminant Anlaysis miscalssified one tomato in the test sample. The analysis by the neural network gives a result without missclassifcation.

4. CONCLUSIONS

The results show that the neural network is suitable for classification. The results of non-destructive and objective methods by the computer vision system is suitable for classification parameters by neural network. The use of neural network has a systematic preliminary work is needed for appropriate selection of training sample. The selection of right shape and color parameters is also very important.

REFERENCES

Székely, V. (1994). *Image correction, sound analysis and space computing by PC*, Computersbooks, Budapest, (In Hungarian)

Martinovich, L. and Felföldi, J. (1996). Measurement of homogenity of onion (*Allium cepa L.*) varieties and lines using computer based shape and colour analyses. *Horticultural Science-Kertészeti tudomány* 1996, 28. (3-4)

Felföldi, J., Firtha, F., and Győri, E. (1994). Application of machine vision for fruit colour evaluation. *Élelmiszerfizikai Közlemények*, 58/2, p. 37-48 (In Hungarian)

Simpson, P. K. (1990). Artificial Neural Systems. Neural Networks: Research and Applications series, Pergamon Press

Firtha, F. (1996). http://physics2.kee.hu/ffirtha/programs/mbpn.htm

FLUORESCENCE BASED QUALITY SORTING OF APPLES USING SELF-ORGANIZING MAPS

Dimitrios Moshou[1], Stijn Wahlen[1], Reto Strasser[2], Ann Schenk[3], Herman Ramon[1]

[1] *K.U.Leuven, Department of Agro-Engineering and Economics, Laboratory for Agro-Machinery and Processing, Kardinaal Mercierlaan 92, Blok E, 3001, Heverlee, BELGIUM, tel: +32-16-321478, fax: +32-16-321994, email: dimitrios.moshou@agr.kuleuven.ac.be*

[2] *University of Geneva, Faculte des sciences, SECTION DE BIOLOGIE, DÉPARTEMENT DE BOTANIQUE ET BIOLOGIE VÉGÉTALE, LABORATOIRE DE BIOÉNERGÉTIQUE 10, ch. des Embrouchis, CH-1254 Jussy-Genève, Switzerland*

[3] *VCBT, Willem De Croylaan 58, 3001, Heverlee, BELGIUM*

Abstract: The chlorophyll fluorescence kinetics of 'Jonagold' and 'Cox' apples, stored under different conditions to induce mealiness, were measured. Three different storage conditions were considered, causing 3 mealiness levels: not mealy, moderately and strongly mealy. Also destructive measurements of the firmness and texture (compression, juiciness and shear) were done. Classification into the different mealiness levels based on the fluorescence measurements was superior to a classification based on the destructive measurements. To estimate the mealiness level in a non-destructive way from the fluorescence features, a number of different classifiers was constructed. Quadratic discriminants and supervised and unsupervised neural networks were tested and compared. The SOM in general outperforms the other methods. The different advantages of the constructed classifiers suggest that fluorescence can be used in an automatic sorting line to assess the mealiness, and the senescence in general, and eventually to predict the time a certain quality level can be guaranteed to the consumers under certain conditions. *Copyright © 2001 IFAC*

Keywords: Neural Networks, Self-Organizing Systems, Classification, Agriculture, Preprocessing, Pattern Recognition, Quality Control

1. INTRODUCTION

With the newest sorting machines it is possible to separate apples according to colour, size and clearly visible damages like hail damage, scab and older bruises, simply and solely based on imaging with colour cameras. However these advanced machines are not able to select on internal quality attributes (like firmness, sugar content, ripeness). There have been proposed a lot of measurement techniques for a non-destructive assessment of firmness in literature, but none of them has been applied yet on sorting machines as a common tool for distinction of different firmness levels. One of the promising techniques is chlorophyll fluorescence, proposed by Song *et al.* (1997) for assessing the firmness of apples. Chlorophyll fluorescence is used to investigate the photosynthetic activity of leaves and to detect a wide variety of stresses in plants. A lot of factors, directly or indirectly acting on the photosynthesis, will influence the chlorophyll fluorescence: light intensity, temperature, humidity, gas composition, senescence, herbicide treatment, species, drought, nutrient shortages, diseases, the 'entire' prehistory of a plant. Changes in fluorescence can be linked to the ageing and ripening, causing decrease of firmness, chlorophyll loss and loss of photosynthetic activity per unit chlorophyll. Combined with advanced neural network based techniques like the SOM, an automated high performance classifier based on fluorescence kinetics response has been constructed.

2. MATERIALS AND METHODS

2.1 Fluorescence principles and techniques

The photosynthesis in plant cells occurs in special cell organelles, the chloroplasts. The entering light energy is absorbed by the light-harvesting complex pigments situated in an internal membrane system (thylakoids), from where the light energy is transported to the reaction centres of photosystem II (PSII) and I (PSI) by resonance energy transfer. This energy turns the reaction centre P680 of photosystem II in an excited state. In this state the reaction centre has a high reducing capacity and loses an electron to pheophitin. The positively charged P680 attracts an electron from the H_2O splicing complex. The electron from the reduced pheophtin is passed through a number of electron acceptors to the reaction centre of PSI, P700. From P 700 the electron is, after excited by light energy from the light-harvesting complex (LHC), finally delivered to $NADP^+$. The reduced form, NADPH, delivers the reducing power necessary for the chemical fixation of CO_2. The light energy absorbed by the LHC is converted into chemical energy (photosynthesis), heat and fluorescence. The fluorescence of green plants is approximately 3-5% of the excitation energy and is almost exclusively emitted by chlorophyll *a*. The level of fluorescence depends on the reduction state of the plastoquinones Q_A and PQ, two of the electron acceptors between PSII and PSI. When the pool of these acceptors is in a highly reduced state, the electron flow, and the related energy transport for the reduction of CO_2 is impeded and a maximum amount of the absorbed energy is remitted as fluorescence. Fluorescence is not only lowered by reoxidation of the quinones (photochemical quenching), but also by other mechanisms (non-photochemical quenching) with the most important ones: (1) acidification of the lumen of the thylakoid; (2) release of a part of the LHC II from PS II after phosphorylation; (3) photo-inhibition of PSII (Krause & Weiss, 1991).

The photosynthesis of plants reaches an equilibrium state or steady state under a certain light condition. In this state the photosynthesis and fluorescence are roughly constant. When a plant under these steady state conditions is suddenly illuminated with a light of stronger intensity, the fluorescence rises to a peak within one second. Then the fluorescence lowers until, after some minutes, a steady state is reached of approximately the same intensity as the previous. Under steady-state conditions the electrons flow at a certain rate from the H_2O splicing complex to $NADP^+$ while picking up energy at the reaction centres. When plants are suddenly illuminated with stronger light, more energy is absorbed by the LHCes and more electrons can be withdrawn from water. But, the rate-limiting process in the electron chain is the delivery of electrons to CO_2, causing a piling up of electrons at the electron acceptors between PSII and PSI. This process brings the plastoquinones Q_A and PQ in a highly reduced form, resulting in a rise of the fluorescence level. After some time the CO_2 reducing part of the electron transport chain start to adapt to the higher light level, the electron flow increases, and the fluorescence diminishes to the steady state level. In short: due to an imbalance between the higher amount of light energy and the slow adapting electron transport rate, the abundance of absorbed energy is visual as a rise in the fluorescence.

The chlorophyll fluorescence in apples follows the same course (rise to a peak when illuminated with a stronger light), but the intensity of the fluorescence is lower because of the lower chlorophyll content of apple fruit.

As earlier mentioned, the chlorophyll fluorescence depends on the prehistory of the plant, including the light intensity before measuring, which turns the plant in a certain light adapted state. For ease of comparing different measurements, to have a reference light situation and to obtain some phenomenological parameters (Strasser and Strasser, 1995), measurements are often done on dark-adapted plants. At dark adaptation, only the duration of the dark adaptation has to be defined, at light adaptation the light intensity and light composition should also be included. Based upon the chlorophyll fluorescence, fluorescence parameters are determined that can be linked to photosynthetic parameters that give a quantitative value for the basic functioning of photosynthetic electron transport. Our purpose now is to use fluorescence to differentiate apples according to firmness based on the measured curves, details about the relationship between fluorescence and photosynthetic parameters can be found in: Strasser *et al.* (1999).

2.2 Fluorescence data collection

The fluorescence was recorded with the PEA (Plant Efficiency Analyser) fluorimeter of Hansatech. The fluorescence is excited by ultra-bright LEDs with a peak wavelength of 650 nm. Chlorophyll fluorescence signals are detected using a PIN photocell after passing through a long pass filter (50% transmission at 720 nm). The recording time during the experiments was 1 sec with a resolution of 10 microsec during the first 2 millisec and after that with a resolution of 1 millisec. The apples were not dark-adapted and the measurements occurred under normal artificial lighting. One part of a leaf clip was put on the measuring head of the fluorometer, limiting the measuring area to a circle of 4mm diameter. The clip was then pushed against the apple, avoiding the entrance of ambient light into the photocell.

2.3 Fruit Samples

The experiments were done on the apple varieties Jonagold and Cox. Different mealiness levels were artificially induced by storing the fruit during different periods under high relative humidity (R.H.) at room temperature. All apples were well acclimatised to room temperature at start of the measurements. The fresh Cox (mealiness level 1) were stored at 3 °C, a O_2 content of 2.2 % and CO_2 content of 0.7 %in the air. The medium mealy Cox (level 2) and the high mealy Cox (level 3) were stored at 20 °C and 95 % R.H. during 10 days and 25 days respectively before measuring. The fresh Jonagold (mealiness level 1) were stored at 1 °C, a O_2 content of 1 % and CO_2 content of 2 % in the air. The medium mealy Jonagold (level 2) and the high mealy Jonagold (level 3) were stored at 20 °C and 95 % R.H. during 10 days and 26 days respectively before measuring. For each mealiness level, 20 apples were measured at the red and green side. Moreover, 20 apples of various unknown mealiness levels, bought in commerce, were also measured.

3. DATA ANALYSIS

3.1 Relation between mealiness level, fluorescence and hardness

In the figures below only the fluorescence value measured after 2 microsec (~ Fo) is plotted against the firmness and the hardness for the different mealiness levels. You can see a better separation by the fluorescence than by the destructive hardness or firmness measurements.

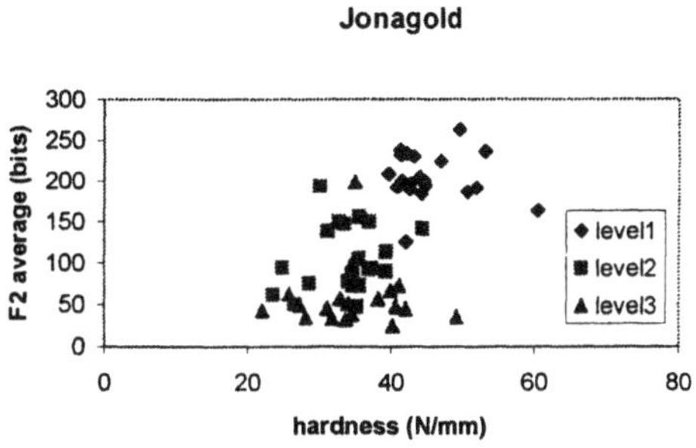

Fig 1 Classification of Jonagold according to different mealiness levels based on hardness and fluorescence F2 average: average fluorescence of one measurement at the red side and one at the green side of the apple.

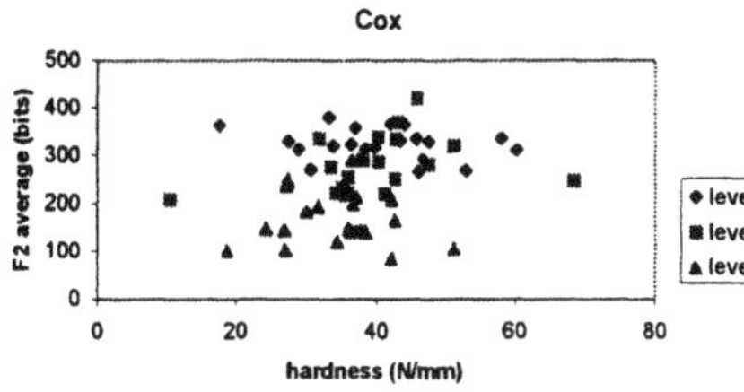

Fig 2 Classification of Cox according to different mealiness levels based on hardness and fluorescence

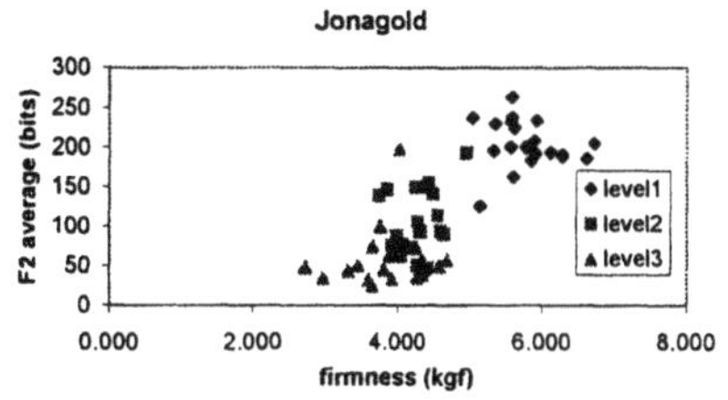

Fig 3 Classification of Jonagold according to different mealiness levels based on firmness and fluorescence. The PCs are based on the whole fluorescence curve.

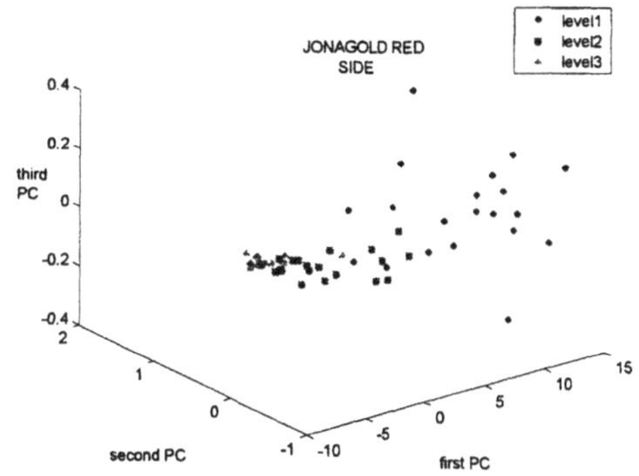

Fig.4 Plot of the first 3 PCs of the fluorescence data measured on Jonagold (red side)

3.2 Classifier construction using Quadratic Discriminant and MLP

The Quadratic Discriminant used was a quadratic combiner trained with an Least Mean Squares (LMS) rule (McLachlan, 1992). Feed-forward neural networks (Bishop, 1995) provide a general framework for representing nonlinear functional mappings between a set of input and output variables. This is achieved by representing the nonlinear function of many variables in terms of composition of nonlinear functions of a single variable, called activation functions. Of course the above notions can be extended to include more than one hidden layers. For the neurons of the feedforward network some commonly used activation functions include the logistic sigmoid and the hyperbolic tangent. In the current paper for reasons of comparison with the PNN classifier, a Multi-Layer Perceptron (Bishop, 1995) was trained with Bayesian Regularisation. The number of hidden neurons was not determined by trial and error as it is usually observed in the bibliography. A sufficiently large number of neurons were selected for the hidden layer (in this case 10). Bayesian Regularisation was used to assure the generalisation performance of the neural network. This procedure is used to avoid overtraining and overfitting. It works well independent of the size of network used and it leaves considerable freedom with respect to the selection of the hidden layer size, since the weight constraints that are imposed are equivalent to regularization or smoothing of the error function used, independent of the complexity of the error surface.

Table 1. Performance matrix of the quadratic discriminant (QD) based prediction of mealiness level for the Jonagold and Cox variety observations using the first 2 principal components from the fluorescence response curve.

	observations classified correctly by QD	
From level	Jonagold	Cox
1	35 (87.5%)	24 (60%)
2	20 (50%)	28 (70%)
3	36 (90%)	36 (90%)

Table 2. Performance matrix of the MLP based prediction of mealiness level for the Jonagold and Cox variety observations using the first 2 principal components from the fluorescence response curve

	observations classified correctly by MLP	
From Level	Jonagold	Cox
1	36 (90%)	33 (82.5%)
2	24 (60%)	18 (45%)
3	34 (85%)	37 (92.5%)

3.3 Classifier construction using a self-organizing Map (SOM)

The Self-Organizing Map (Kohonen, 1995) is a neural network (NN) that maps signals ($\mathbf{x}$) from a high-dimensional space to a one- or two-dimensional discrete lattice of neuron units ($\mathbf{s}$). Each neuron stores a weight ($\mathbf{w}_s$). The map preserves topological relationships between inputs in a way that neighbouring inputs in the input space are mapped to neighbouring neurons in the map space. The learning algorithm for the input weights is based on the original algorithm of Kohonen and is formulated in equation:

$$\Delta \mathbf{w}_s^{(in)} = \varepsilon h(\mathbf{x} - \mathbf{w}_s^{(in)}) \qquad (1)$$

Where ε and h are the learning rate and the Gaussian neighborhood kernel respectively.

A way of using the SOM to find correlations between the data is to label the neurons of the SOM using a different set than the training set and finding the best-matching-units (BMUs) for every example in the testing set or labelling set. Some of the neurons that are selected most frequently by examples of one class are labelled based on a voting procedure. These neurons are then able to estimate the class of a new example presented to the SOM by calculating the Euclidean distance of the example vector to the codebook vector of each neuron and finding the BMU. The label of the BMU is then the estimated class of the new example vector. Applying a 10x10 SOM to the Jonagold and Cox observations gives very high separation compared to a quadratic discriminant and the MLP. The SOM results are presented in the tables 3 and 4 while figures 5 and 6 show the labelled maps. From the shape of the clusters it is evident that the different classes are not well separated. The neurons that are not labelled constitute the borders of the classes and show the degree at which example observations from one class can be misclassified as observations belonging to another class.

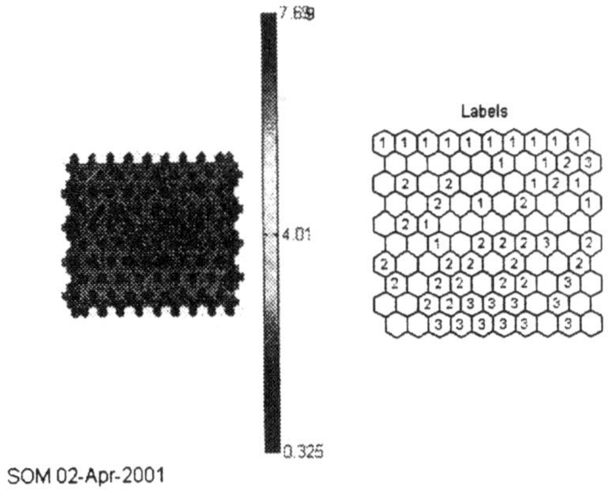

Figure 5. The SOM used to classify the Cox variety fluorescence data with the labels representing mealiness classes.

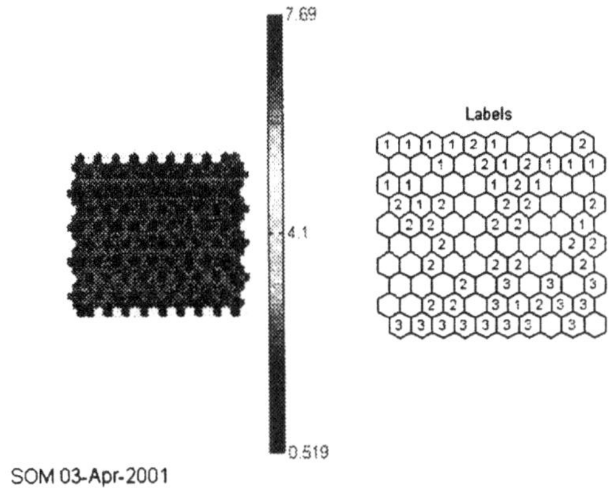

Figure 6. The SOM used to classify the Jonagold variety fluorescence data with the labels representing mealiness classes

Table 3. Confusion matrix of the SOM based prediction of mealiness level for the Jonagold variety observations using the first 2 principal components from the fluorescence response curve.

	Jonagold		
	observations classified into		
from	level 1	level 2	level 3
level 1	39 (97.5%)	0 (0%)	1 (2.5%)
level 2	1 (2.5%)	33 (82.5%)	6 (15%)
level 3	0 (0%)	4 (10%)	36 (90%)

Table 4. Confusion matrix of the SOM based prediction of mealiness level for the Cox variety observations using the first 2 principal components from the fluorescence response curve.

	Cox		
	observations classified into		
from	level 1	level 2	level 3
level 1	31 (77.5%)	8 (20%)	1 (2.5%)
level 2	2 (5%)	34 (85%)	4 (10%)
level 3	0 (0%)	2 (5%)	38 (95%)

4. CONCLUSIONS

Chlorophyll fluorescence can be an attribute to an advanced quality sorting of apple fruit due to its speed and non-destructive character, but there still remain some problems to be solved. There was an effort to link the chlorophyll fluorescence with the mealiness of apple fruit, supposing mealiness and chlorophyll degradation go hand in hand, but also senescence causes chlorophyll degradation and changes in the chlorophyll fluorescence. The influence of ageing is not taken into account in this study, but cannot be neglected in practice. After all, mealiness is nothing else than an (accelerated) senescence symptom.

In this study, a difference in the general fluorescence level was found for the two varieties, so one can expect also differences for other varieties. The use of advanced data mining techniques like the SOM offers a clear advantage over conventional neural network and discriminant techniques in the automated sorting of fruits based on non-destructive techniques and especially fluorescence based features.

REFERENCES

Bishop, C. M. (1995) *Neural Networks for Pattern Recognition*. Oxford University Press, Oxford.

Johnson, R. A. & Wichern, D. W. (1998). *Applied multivariate statistical analysis*. Prentice Hall, London. 816p.

Kohonen, T. (1995). *Self-Organizing Maps*. Springer Series in Information Sciences.

Krause, G. H. & Weis, E. (1991). Chlorophyll fluorescence and photosynthesis: the basics. *Annual Review of Plant Physiology and Plant Molecular Biology*, 42: 313-349.

McLachlan, G. J. (1992). *Discriminant Analysis and Statistical Pattern Recognition.* John Wiley, New York

SAS Institute Inc.(1989). SAS/STAT® User's Guide Version 6, Fourth Edition, Volume 2, Cary, NC: SAS Institute Inc. 846 pp.

Song, J., Deng, W.& Beaundry, R. M. (1997). Changes in chlorophyll fluorescence of apple fruit during maturation, ripening, and senescence. *HortScience*, 32: 891-896.

Strasser, B. J. & Strasser, R. J. (1995). Measuring fast fluorescence transients to address environmental questions: The JIP test. In Mathis, P. (Ed.) *Photosynthesis from Light to Biosphere*, vol V, pp. 977-980, Kluwer Academic Publisher, the Netherlands.

Strasser, R. J., Strivastava, A. & Tsimilli-Michael, M. (1999). The fluorescence as transient tool to characterize and screen photosynthetic samples. In Yunus; M., Pathre, U. & Mohanty, P. (Eds.) *Probing photosynthesis: mechanism, regulation and adaptation.* Taylor and Francis, London.

DEVELOPMENT OF AN AUTOMATIC RICE-QUALITY INSPECTION SYSTEM

Shuso Kawamura[1], Motoyasu Natsuga[2] and Kazuhiko Itoh[1]

[1]*Agricultural Process Engineering Laboratory, Graduate School of Agricultural Science, Hokkaido University, Sapporo 060-8589, Japan.*
Email, shuso@bpe.agr.hokudai.ac.jp
[2]*Agricultural Ecology and Engineering Laboratory, Faculty of Agricultural Science, Yamagata University, Tsuruoka 997-8555, Japan*

Abstract: The need has arisen in rice-drying facilities in Japan for an accurate method to measure quality aspects of rice when it arrives at the drying facility. The precision and accuracy of a near-infrared (NIR) spectrometer and a visible light (VIS) segregator were found to be sufficiently high to determine moisture content, protein content and sound whole kernel rate. An automatic rice-quality inspection system was consequently developed. The system consists of a rice huller, a rice cleaner, an NIR spectrometer and a VIS segregator. Based on rice-quality information, this system enables rough rice transported to a rice-drying facility to be classified into six qualitative grades.

Keywords: Spectroscopy, Quality, Accuracy, Precision, Validation, Classification

1. INTRODUCTION

In Japan, harvested rough rice is first transported to rice-drying facilities, where the moisture content of the rough rice is checked. After moisture inspection, the damp rough rice is dried up to a moisture content level of about 15% (wet basis), and then the hulls are removed. Brown rice (i.e., rice from which the hulls have been removed) is then transported to rice-milling facilities, where the bran layers and embryos are removed. Milled rice is usually used for cooking.

Components of brown rice, such as sound whole kernels, immature kernels and underdeveloped kernels, are important for assessing the quality of rice. An inspector from the Japan Food Agency usually visually evaluates the components of the rice as an official inspection.

The major chemical constituents of milled rice are moisture (15%), protein (7%) and starch (87%). The protein content of milled rice is a very important quality aspect, especially in East Asian countries, where people eat short-grain, non-waxy rice.
Ishima *et al.* (1974), Yanase *et al.* (1984), and Shibuya (1990) reported that the protein content of rice is important for the following reasons. Protein inhibits water absorption and starch swelling when milled rice is cooked, and it greatly affects the texture of cooked rice. Rice with a high protein content is less sticky when cooked. Since East Asian people prefer sticky cooked rice, rice with a low protein content is preferable in East Asian countries.
Chikubu et al. (1985) reported that protein content and some other physical properties of rice contribute to about 60% of the eating quality.

The need has recently arisen in rice-drying facilities in Japan for an accurate method to measure not only moisture content but also other quality aspects of rice in order to grade rough rice according to quality when it arrives at the drying facility.
The objective of this study was to develop an automatic rice-quality inspection system using a near-infrared spectrometer and a visible light segregator.

2. MATERIALS AND METHODS

2.1 Rice Samples

A popular Japanese short-grain, non-waxy variety of rice, "Kirara 397", was used for samples in this study.

2.2 Near-Infrared Spectrometer

A near-infrared (NIR) spectrometer (model Grainspec-1000J, Foss Electric, Hillerod, Denmark; Figure 1) was used to obtain NIR spectra of damp whole-grain rough rice and damp whole-grain brown rice. The spectrometer could scan 33 wavelengths between 825 nm and 1,075 nm. Damp rough rice and damp brown rice were poured into the hopper of the spectrometer and loaded automatically into a sample cell. The NIR energy that was transmitted through the grain sample was detected by a silicon photocell array, converted into an electrical signal, and processed by a computer to be transformed to log (1/T). One scan for each loading was saved in computer memory. Ten scans were performed, and the data were averaged to obtain an NIR spectrum for each sample. It took about two minutes to perform ten scans for one sample. The grain capacity in the sample cell was adjustable according to the type of grain. This capacity was referred to as the light path length of the cell. After a preliminary test, spectral data for damp rough rice were then collected with a light path length of 25 mm, and spectral data for damp brown rice were collected with a light path length of 30 mm.

Moisture contents of damp rough rice samples and damp brown rice samples were determined by the official method of the Japan Food Agency. Five grams of ground rice were placed in a forced air oven at 105°C for 5 hours, and moisture was computed on a wet basis. The damp rough rice grains were dried up to 15% of moisture content to obtain dried rough rice samples, and the rough rice grains were hulled to obtain brown rice samples. The brown rice grains were then milled to obtain milled rice samples. Protein contents of the brown rice samples and milled rice samples were determined by the Kjeldahl method (N×5.95) and calculated on a dry basis. Table 1 shows the precision of the chemical analysis. The standard deviations of differences among repetitions (SDD) in measured moisture content of damp rough rice (0.32%) and damp brown rice (0.29%) were larger than that of dried rice (Natsuga et al., 1992; Kawamura et al., 2001). The SDD data indicated that high-moisture rice had a large moisture deviation.

Calibration models used to estimate the moisture contents of damp rough rice and damp brown rice were developed from damp rough rice NIR spectra and damp brown rice NIR spectra, respectively, and models used to estimate the protein contents of brown rice and milled rice were also developed from damp rough rice NIR spectra and damp brown rice NIR spectra, respectively. Spectral data analysis software (Data Tracker, Foss Electric, Hillerod, Denmark) was used for chemometric analysis. The samples were randomly divided into two groups: a calibration set containing two thirds of the samples and a validation set containing the remainder (one third) of the samples. The method of partial least squares (PLS) was used to develop calibration models from the original spectra sets. Pretreatment such as smoothing or derivation was not performed on the original spectra.

2.3 Visible Light Segregator

A visible light (VIS) segregator (Model RA60A, Satake Engineering, Tokyo, Japan; Figure 2) was used to determine components of brown rice. Brown rice was poured onto a rotating disk that had oval holes for carrying each rice kernel. A green light sensor and a red light sensor detected the light reflectance from each kernel. A transmission sensor detected interior characteristics of each kernel. Injection nozzles under the rotating disk were used to separate kernels into sound whole kernels, immature kernels, underdeveloped kernels, damaged kernels and discolored kernels. It took about one minute to separate 1000 brown rice kernels into these five types.

Table 1 Precision of chemical analysis

Chemical analysis	Sample	n[1]	Range (%)	SDD[2] (%)
Moisture content	Damp rough rice	134	15.4 - 34.8	0.32
	Damp brown rice	136	15.9 - 32.1	0.29
Protein content	Brown rice	150	7.3 - 9.4	0.07
	Milled rice	150	6.7 - 8.9	0.05

[1]Number of samples.

[2]Standard deviation of differences among repetitions.

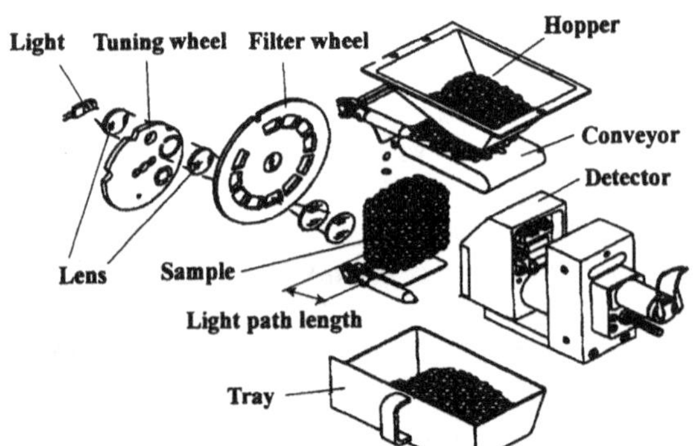

Fig. 1. Near-infrared spectrometer.

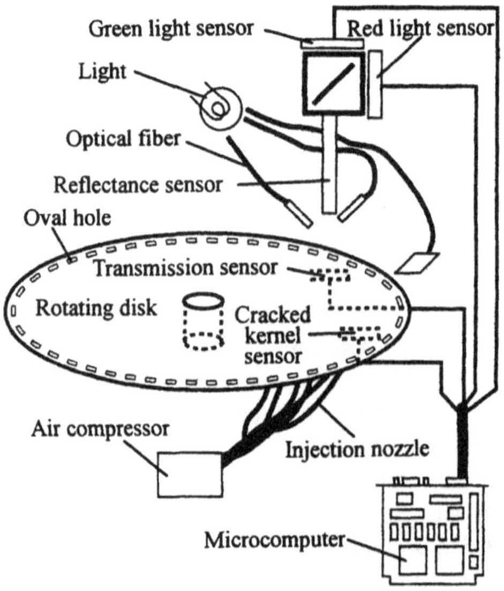

Fig. 2. Visible light segregator.

As a reference, human visual inspection was performed according to the official inspection standards of the Japan Food Agency. Table 2 shows the precision of the reference analysis. The value of SDD in measured sound whole kernels (0.44%) was larger than that of discolored kernels (0.02%). The SDD data indicated that the component with a high percentage had a large deviation.

3. RESULTS AND DISCUSSION

3.1 Results of Analysis Using a Near-Infrared Spectrometer

The results of PLS calibration modeling and validation statistics for moisture content and protein content are summarized in Table 3. The coefficients of determination (r^2) of the validation set were sufficiently high (0.96 and 0.97) for determination of moisture content. The standard errors of prediction (SEP) for moisture content (0.70% and 0.50%) were slightly worse than those previously reported (Natsuga et al., 1992; Kawamura et al., 2001), because damp rough rice and damp brown rice had large moisture deviations. However, the results shown in Figure 3 indicated that NIR spectroscopy could be used for accurate determination of the moisture content of damp whole-grain rough.

Table 2 Precision of reference analysis

Reference analysis	n	Range (%)	SDD (%)
Sound whole kernel	9	59.9 - 90.0	0.44
Immature kernel	9	8.6 - 27.8	0.43
Underdeveloped kernel	9	0.1 - 6.1	0.15
Damaged kernel	9	1.3 - 5.9	0.18
Discolored kernel	9	0.0 - 0.4	0.02

Abbreviations as in Table 1.

Brown rice protein content and milled rice protein content were determined using calibration models obtained from damp rough rice spectra and damp brown rice spectra, respectively (Table 3). Scatter plots of chemical analysis vs NIR estimated values of the protein content of brown rice and that of milled rice are shown in Figures 4 and 5, respectively. The r^2 values of 0.70 and 0.76 and the SEP values of 0.24% and 0.22% were worse than those previously obtained from dried ground brown rice (r^2=0.98 and SEP=0.15%, Suzuki et al., 1996), dried ground milled rice (r^2=0.94 and SEP=0.17%, Natsuga et al., 1992) and dried whole-grain milled rice (r^2=0.97 and SEP=0.13%, Delwiche et al., 1996). In previous studies, the protein contents of brown rice and milled rice were determined from their spectra, while the protein contents of brown rice and milled rice in this study were determined from damp rough rice spectra. The accuracy of the calibration models in this study was, therefore, lower than that in previous studies.

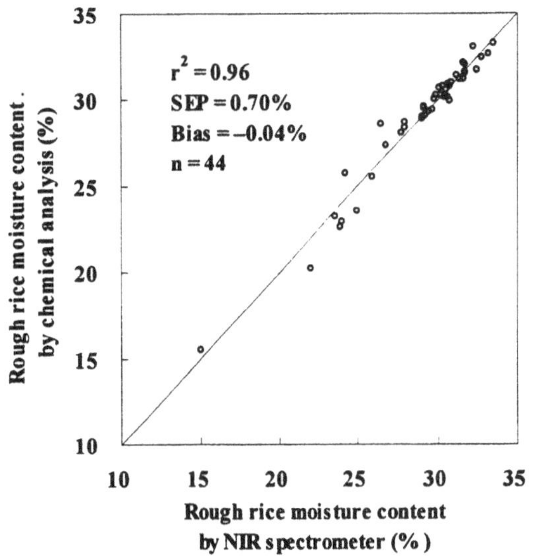

Fig. 3. Validation of the NIR spectrometer for determining moisture content of damp rough rice.

Table 3 Calibration and validation statistics for moisture content, brown rice protein content and milled rice protein content

Constituent content	Sample measured NIR spectra	Calibration model					Validation				
		n[1]	Range (%)	nF[2]	r^{2} [3]	SEC[4] (%)	n[1]	Range (%)	r^{2} [3]	SEP[5] (%)	Bias (%)
Moisture content	Damp rough rice	90	15.4 - 34.8	7	0.97	0.70	44	15.5 - 33.2	0.96	0.70	-0.04
	Damp brown rice	96	15.9 - 32.0	8	0.99	0.37	40	16.5 - 32.1	0.97	0.50	0.06
Brown rice protein content	Damp rough rice	100	7.3 - 9.4	11	0.86	0.18	48	7.4 - 9.3	0.70	0.24	0.04
	Damp brown rice	100	7.3 - 9.4	7	0.74	0.23	48	7.4 - 9.3	0.68	0.23	0.02
Milled rice protein content	Damp rough rice	100	6.7 - 8.9	12	0.74	0.26	49	6.7 - 8.7	0.76	0.22	-0.05
	Damp brown rice	99	6.7 - 8.9	7	0.74	0.24	49	6.7 - 8.7	0.71	0.25	0.02

[1]Number of samples. [2]Number of factors. [3]Coefficient of determination.
[4]Standard error of calibration. [5]Standard error of prediction.

The results shown in Figures 4 and 5, however, indicate that an NIR spectrometer has a reasonable ability to classify damp rough rice into high, middle and low protein content groups upon arrival at a drying facility. Shenk and Westerhaus (1993) investigated the relationship between r^2 and the percentage of times a sample is correctly classified, and they reported that about 70% of the samples could be correctly classified if damp rough rice samples are divided into three protein content groups by using the calibration model shown in Figure 5 (r^2=0.76).

3.2 Results of Analysis using a Visible Light Segregator

The results of validation statistics for component analysis using a VIS segregator are summarized in Table 4.

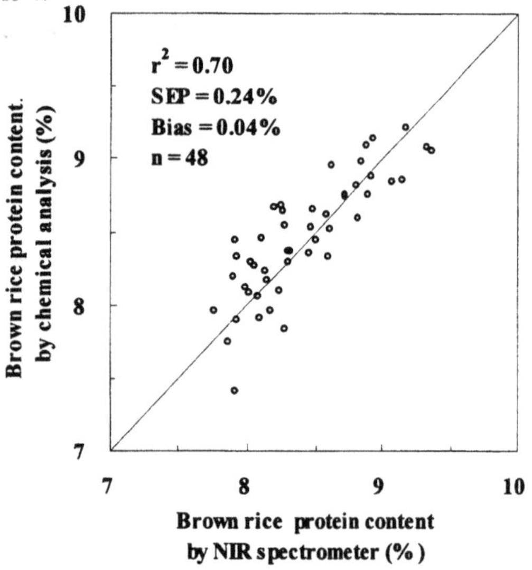

Fig. 4. Validation of the NIR spectrometer for determining protein content of brown rice using a calibration model obtained from damp rough rice spectra.

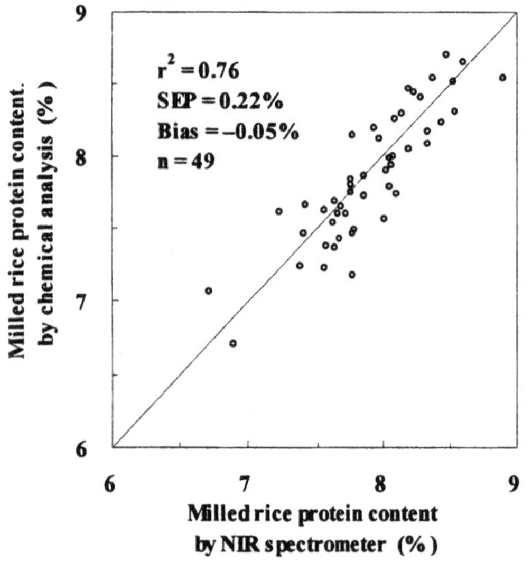

Fig. 5. Validation of the NIR spectrometer for determining protein content of milled rice using a calibration model obtained from damp rough rice spectra.

Scatter plots of results of human inspection vs segregator measured values of sound whole kernels and immature kernels are shown in Figures 6 and 7, respectively. The r^2 values of 1.00 and 0.99 and the SEP values of 1.34% and 0.84% are sufficiently high, indicating that a VIS segregator could be used instead of human visual inspection for component analysis of brown rice.

<u>Table 4 Validation statistics for component analysis</u>

Component	n	r^2	SEP (%)	Bias (%)
Sound whole kernel	9	1.00	1.34	-2.02
Immature kernel	9	0.99	0.84	1.70
Underdeveloped kernel	9	0.99	0.25	-0.29
Damaged kernel	9	0.86	0.66	0.70
Discolored kernel	9	0.23	0.17	-0.10

Abbreviations as in Table 3.

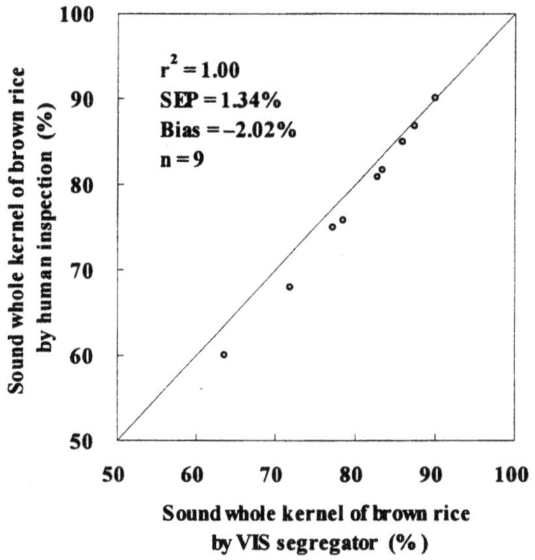

Fig. 6. Validation of the VIS segregator for determining sound whole kernels of brown rice.

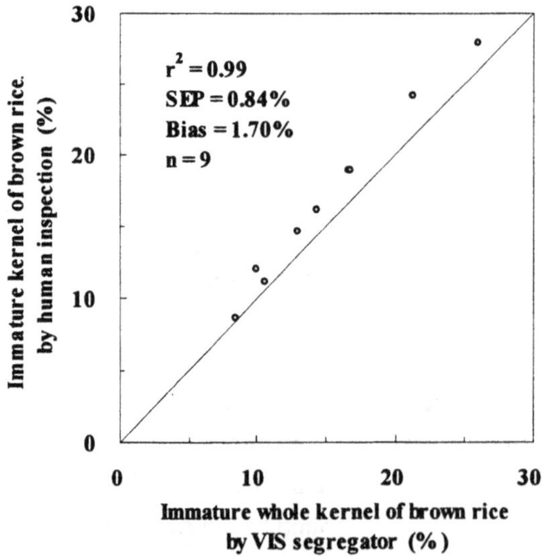

Fig. 7. Validation of the VIS segregator for determining immature kernels of brown rice.

3.3 Design of an Automatic Rice-Quality Inspection System

The precision and accuracy of the NIR spectrometer and the VIS segregator were sufficiently high to enable accurate classification of rice arrived at a drying facility. Based on the results, an automatic rice-quality inspection system was designed. As shown in Figure 8, the system consisted of a rice huller (an impeller-type huller), a rice cleaner (a thickness grader), an NIR spectrometer and a VIS segregator. Each rice sample was moved automatically from one apparatus to the next one through tubes by pneumatic conveyors, bucket elevators or the force of gravity. Rough rice samples should be hulled so as not to become stuck in the tubes or in one of the apparatuses and also to enable measurement of sound whole kernels of brown rice. Figure 9 shows a picture of an inspection room equipped with an NIR spectrometer, a VIS segregator and a computer. The rice sample passes through a huller and a rice cleaner in a room above the inspection room, and the brown rice drops down through the ceiling into the inspection room. The computer controls all apparatuses and receives information from each apparatus. Based on information of quality aspects (protein content, moisture content and sound whole kernel rate), rough rice transported to a rice-drying facility can be classified into six qualitative grades (three protein content levels times two sound whole kernel rates). About two hundred automatic rice-quality inspection systems are currently being used at rice-drying facilities in Hokkaido, the northernmost island of Japan.

4. CONCLUSIONS

The precision and accuracy for determining protein content, moisture content and sound whole kernel rate using an NIR spectrometer and a VIS segregator were sufficiently high to enable classification of rice samples. An automatic rice-quality inspection system was designed. The system consists of a rice huller, a rice cleaner, an NIR spectrometer and a VIS segregator. In this system, a rice sample is moved automatically from one apparatus to the next one. Based on the information of the quality aspects, this system enables rough rice transported to a rice-drying facility to be classified into six qualitative grades.

REFERENCES

Chikubu, S., S. Watanabe, T. Sugimoto, N. Manabe, F. Sakai and Y. Taniguchi (1985). Establishment of palatability estimation formula of rice by multiple regression analysis. *J. of the Japanese Society of Starch Science,* **32**, 51-60.

Delwiche, S.R., K.S. McKenzie and B.D. Webb (1996). Quality characteristics in rice by near-infrared reflectance analysis of whole-grain milled samples. *Cereal Chem.,* **73**, 257-263.

Ishima, T., Hi. Taira, Ha. Taira and K. Mikoshiba (1974). Effect of nitrogenous fertilizer application and protein content in milled rice on olganoleptic quality of cooked rice. *Report of National Food Research Institute, Japan,* **29**, 9-15.

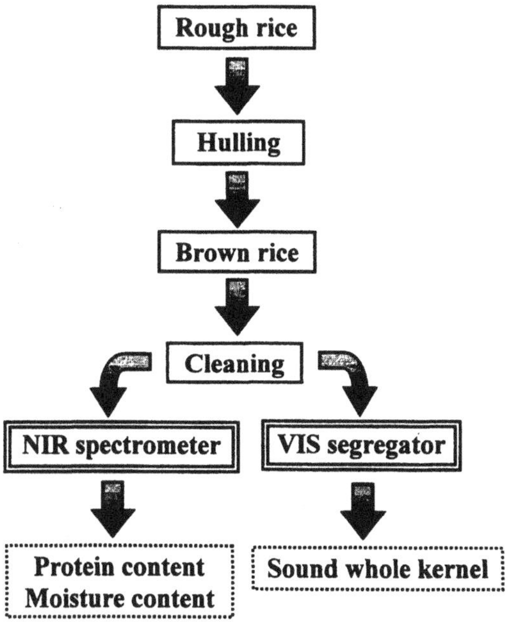

Fig. 8. Automatic rice-quality inspection system.

Fig. 9. Picture of an inspection room equipped with an NIR spectrometer, a VIS segregator and a computer.

Kawamura, S., K. Takekura and K. Itoh (2001). Accuracy in determination of rice constituent contents using near-infrared transmission spectroscopy and improvement in the accuracy. *J. of the Japanese Society of Agricultural Machinery,* in press.

Natsuga, M., S. Kawamura and K. Itoh (1992). Precision and accuracy of near-infrared reflectance spectroscopy in determining constituent content of grain. *J. of the Japanese Society of Agricultural Machinery,* **54**, 89-94.

Suzuki, Y., S. Takahashi, M. Takebe and K. Komae (1996). Factors that affect protein contents in brown rice estimated by a near-infrared spectrometer. *Nippon Shokuhin Kogyo Gakkaishi,* **43,** 203-210.

Shenk, J.S. and M.O. Westerhaus (1993). *Analysis of agriculture and food products by near-infrared reflectance spectroscopy. Part 1,* 13-16. Infrasoft International Co. Port Matilda, Pa.

Shibuya, N. (1990). Chemical structure of cell walls of rice grain and grain quality. *Nippon Shokuhin Kogyo Gakkaishi* **37**, 740-748.

Yanase, H., K. Ohtsubo, K. Hashimoto, H. Sato and T. Teranishi (1984). Correlation between protein contents of brown rice and textural parameters of cooked rice and cooking quality of rice. *Report of National Food Research Institute, Japan,* **45**, 118-122.

COMPUTER AIDED MOISTURE MEASUREMENTS DURING FRUIT DRYING PROCESSES

István Seres[1], István Farkas[1], László Font[2]

[1]*Department of Physics and Process Control, Szent István University Gödöllö, Páter K. u. 1., H-2103 Hungary*

[2] *Process Control Research Group, Hungarian Academy of Sciences SZIUG Páter K. u. 1., Göddöllő, H-2103 Hungary*

Abstract: This paper deals with the experiences concerning to fruit drying controlled by computer aided image analyis process. Two type of fruits were studied as apple slices and blackthorn. First, separate physically based models for both fruits were carried out to describe the heat and mass transfer during the drying process and to determine the diffusion coefficient, as well. In order to validate the model developed measuring tests were performed. *Copyright © 2001 IFAC*

Keywords: drying, fruit, moisture measurement, quality control, computer aided image analysis

1. INTRODUCTION

The technological energy consumption still has a great importance in the agriculture as because of the growing prices and the importance of the environment protection. At the same time the quality control and quality preservation becomes also more and more important items for processing of agricultural products. A traditional and very widely used product preservation is the drying. Main drawbacks of drying are the relatively high energy consumption and changing product properties during the heat and de-watering treatment. However, the attractiveness of drying methods can be improved by using advanced control and optimization techniques for reducing the energy consumption. These methods can only be applied if sufficient information is collected on the interaction between the drying conditions and the change of product properties. The heat and mass transfer properties are crucial for the proper modeling but generally they are not as well defined due to the several number of variables (Whitelock, 1999). However a lot of different model for the physical properties can be found in the mathematical modeling of the drying (Sun, 1995).

Among the different quality properties the appearance is a definitive factor in the moment of the selling/buying especially for agricultural products e.g. fruits. This is the reason to use the image analysis technique for the quality control measurements as an important method. By this way both an overall quality (Murakami, 1994) and some particularly important factors as e.g. the area or volume (Do, 1997) or the shrinkage (Raghavan, 1999) can be estimated.

The dryer is one of the biggest energy consumers in a farm and the costs of energy are still a reasonable factor in overall evaluation of agricultural dryers. No doubt that recently the quality issues of the dried end-product are also key-questions in connection to market-oriented approach. These are the motivation for a multi-objective optimization problem of the drying along with a constraint of the processing time limitation. For quite many of the traditional agricultural drying processes a low temperature technology can be applied, i.e. for fruits, herbs, seeds, mushroom, grass, etc. This fact, along with the recently underlined environmental questions, also implies reconsidering the application of solar energy in drying.

Nowadays for instance in Hungary, instead of the former large-scale farms there are numerous small farms operated by families. These small farms need some simple easy-to-operate type of agricultural equipment, which do not require too much investment and maintenance but can be operated economically.

2. THE DRYING EQUIPMENT

On the basis of the explanations given before a small size dryer operated by solar energy seems to be realistic in many reasons. Such a dryer can cover the drying demand of a single farm. It is not planned to use as a raw corn drying or to replace the large-volume oil and gas heating dryers used from the mass grain production. Taking into account the high energy prices, environmental considerations and that the equipment has to operate in the fields far from the electrical grid the use of solar energy was planned for the artificial ventilation.

The solar dryer planned for such purposes has three main parts:

- A dryer (drying cabin) with different trashes for the different products.
- A PV module with an electrical fan for artificial air circulation.
- An air solar collector is attachable to the dryer for preheating the inlet air.

The adjustment of the different parts can be seen in Fig. 1.

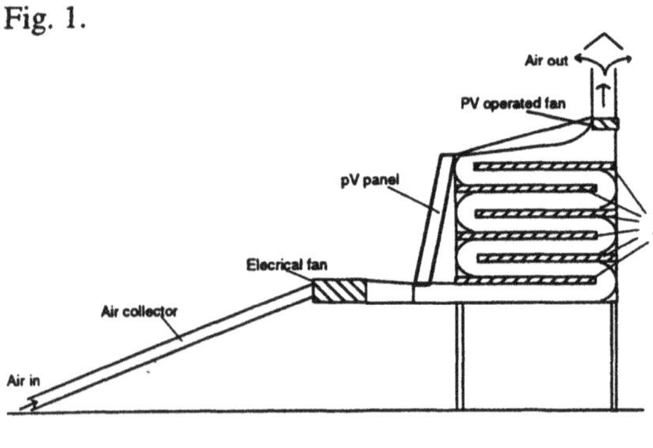

Fig. 1. Scheme of the modular solar dryer

Because of the modular construction of the dryer it can be operated in different modes:

- Natural ventilation of ambient air with a chimney.
- Artificial ventilation of ambient air when the PV module is applied.
- Artificial ventilation of the drying air preheated by a solar air collector.
- The combination of the above modes can also be used.

More information about the dryer can be get from earlier publications (Farkas et. al, 1996).

3. SELECTION OF THE FRUITS TO BE DRIED

During the selection of the fruits to be studied different aspects, were considered. Beside the popularity of the different fruits the importance of the drying for the given fruit were also considered. Another aspect was to find some food industrial or agricultural partner for whom the given fruit is important in the sense of marketizing.

Among the usually dried products, in Hungary, one of the most popular fruit is the apple. It is dried as a basic component of fruit tee or taste additive for different foods (muesli, etc.) or just for preserving. Additionally, several other fruits were considered e.g. grape, peach, plum, cherry and sour cherry but, finally, a not so wide-spread forest fruit, the blackthorn was chosen. The reason for the decision was that an industrial partner showed interested in such kind of study with blackthorn and they supplied also the necessary raw material.

4. MODELING OF THE MOISTURE CONTENT CHANGE

Before the measurements a modeling of the drying process was carried out. As the behavior of the two examined products (apple, blackthorn) is quite different during the drying they were studied separately.

4.1 The modeling for apple

As the apple is dried usually in parts, therefore first the drying behavior of apple slices were investigated in order to get the physical quantities, for instance, the diffusion coefficient. The slice was considered to be a plate and the diffusion coefficient of the apple pulp was assumed much bigger than the diffusion coefficient of the skin so the direction of the moisture flow was considered to be at right angles to the plate. The partial differential equation system of the drying process was set up as follows:

The mass balance of the apple (diffusion equation):

$$\frac{\partial C}{\partial t} = \mathrm{div}(D\,\mathrm{grad}C). \tag{1}$$

The mass balance of the drying air:

$$G\frac{\partial H}{\partial x} = -\rho_b \frac{\partial M}{\partial t}. \tag{2}$$

The energy balance of the apple slice:

$$\rho_b\left(c_p + c_w M\right)\frac{\partial \Theta}{\partial t} = \rho_b \Delta H(\Theta)\frac{\partial M}{\partial t} + U(T-\Theta). \quad (3)$$

The energy balance of the air:

$$\rho_a\left(c_a + c_w H\right)\frac{\partial T}{\partial x} = \rho_b \frac{\partial M}{\partial t} c_w (T-\Theta) - U(T-\Theta). \quad (4)$$

The partial differential is much too complicated to find an analytical solution so it was solved numerically by the finite difference method with the help of the Matlab software.

4.2 The modeling for blackthorn

The above mentioned methods are suitable only for the homogenous systems e.g. for the pulp without skin. This was the reason why another model was developed for the blackthorn.

The resistance of a spherical system for the diffusion in radial direction is :

$$\Re = \frac{1}{D 4\pi}\frac{R-r}{R r}, \quad (5)$$

where R and r is the outer and inner radius of the product and D is the diffusion coefficient.

This formula is valid for the inside of the fruit (for the pulp). The skin was considered as a thin plain because of its thickness. The resistance of the skin can be calculated as :

$$\Re_p = \frac{d}{D_p 4R^2\pi}, \quad (6)$$

d is the thickness of the skin, R is its average radius and D_p is the diffusion coefficient.

So, finally the total resistance of the blackthorn is:

$$\Re_t = \frac{d}{D_p 4R^2\pi} + \frac{1}{D 4\pi}\frac{R-r}{R r}. \quad (7)$$

During the drying the capacity of the system is the volume (V), so the time constant of the process is

$$\tau = \Re_t V = \left(\frac{d}{D_p 4R^2\pi} + \frac{1}{D 4\pi}\frac{R-r}{R r}\right)\frac{4}{3}\left(R^3 - r^3\right)\pi$$

$$\tau = \left(\frac{d}{D_p R^2} + \frac{R-r}{D R r}\right)\frac{\left(R^3 - r^3\right)}{3} \quad (8)$$

Knowing the time constant, the average fruit moisture content against the time can be calculated by the following equation:

$$C(t) = Ce + (C_0 - Ce)e^{-\frac{t}{\tau}}. \quad (9)$$

5. MOISTURE CONTENT MEASUREMENTS

The moisture content of the fruit materials to be dried was measured by traditional way, but the neutron radiography is also planned to be used. During the traditional measurement series mass measurement was performed. At the end of the measurement series the moisture content was determined by a Sartorius M30 fast moisture analyzer. From this moisture content measurement the actual moisture content vs time function was calculated using the measured mass data. These kind of measurements were carried out with both of the studied products.

6. THE IMAGE ANALYSIS

6.1 The used equipment and measurement method

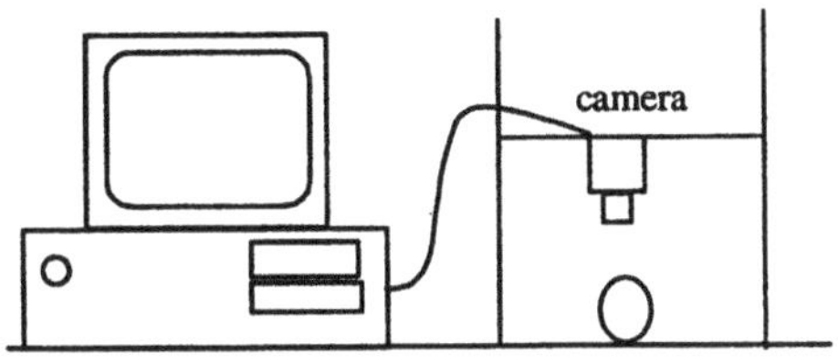

Fig 2. Schematic picture of the measuring system

The scheme of the measuring set-up is shown in Fig. 2. The image analysis was carried out by using a PC camera (type USRobotics, focus: 1 cm - ∞). The images were recorded by the Asymetric Video Capture program which can record still pictures and movies as well. Because of the low speed of the drying process, the pictures were recorded regular times. Beside of the PC camera another digital camera type of Sony Mavica MVC-FD-83 was used for picturing the processes, also. This digital camera record digital photos directly to a 3,5" floppy disk.

The measurements were carried out in two different places, in the drying cabinet of the solar dryer and in an indoor drying cabinet in which controlled environmental circumstances were possible to keep. In the solar dryer cabinet the light conditions were varying a lot depending on the actual weather conditions and the time of the measurement (depending in the part of the day). For this reason, during taking images the fruits were taken to a constant light conditions. As the measurement time was only a few minutes, it has negligible effect on the drying process itself. This problem does not arise with the inside measurements because of the use of artificial lighting.

The images were taken from four different point of views. The fruits - which were held on small trays each time - were photographed together with the tray on a scale in such a way that the screen of the scale is readable on the picture. For the correct size measurements a tape measure were photographed together with the products each time above the scale screen. In this way changing of the mass was recorded each time for the actual moisture content calculations. After that the camera was moved above the tray to take a picture on the products from vertical position. Such kind of pictures were used for the calculation of changing the surface of the blackthorn. Two other pictures were taken from a single product with white and scaled background for the size and color measurements. The recorded pictures were saved to a computer in 640x480 resolution in 24 bit color bitmap format.

6.2 The used methods for the image analysis

The recorded pictures were analyzed by computer program coded in C++. The routines of the program can calculate the area, the contour of the products at right angles to the direction of the image taking, and for the approximately volume of the blackthorn was calculated. These calculations were done with different image analyzing methods, as follows:

- Color $\rightarrow$ gray scale conversion,
- Histogram,
- Segmentation,
- Outline, area and volume calculation,
- Contour search.

The main method of the calculations was as follows. First the color picture was converted to gray scale and a histogram was built up. On the histogram there are two different peaks one is the product the other is the background. The border between them is the depression, which can be read from the histogram. The pixels for what the gray is less then the value for the depression belong to the fruit, so the area at right angles to image can be get from the number of the pixels. From the resolution and the saved image of a tape measure the number of pixels can be converted to length and area.

For the calculation of the outline a contour finder algorithm had to be set up first. This algorithm determines the so-called chain-code. The chain-code shows that the contour on which points goes through. The chain code is calculated from every step from the gray scale of the surroundings pixels. The length of the chain code gives the length of the outline in pixels i.e. in the length of it.

For the blackthorn measurements the volume of the products were calculated, too. In this case the volume was considered to be a sphere with an average radius. From the chain-code the outline of the product can be drawn and an inner and an outer circle can be calculated. The average radius is obviously assumed from the common center of the inner and outer circle. From that value the average volume of the blackthorn can be calculated.

Beside changing in the size another aspects i.e. changing of the surface smoothness and arising of wrinkles were examined. First, it was a trial to calculate the change in the length of the outline, but it was not really successful (Fig. 3). The reason for the failure was that there are two different effects, the decreasing of the size and the increasing of the wrinkles. The measurement of the wrinkles finally was calculated as the ratio of the radius of the outer and inner circles. As the smoothness of the surface goes wrong the difference between the two radiuses increases and the ratio is increasing.

7. RESULTS

For the apple drying the physically based partial differential equation system was set up and transformed to a numerical model solved by the finite difference technique. The numerical model was set-up and tested by the Matlab software. The next step of the research is the determination of the key physical properties of the apple, e.g. the diffusion coefficient and the heat transfer coefficient. For this purpose DNR measurements are also planned to use. From the time dependent moisture content distribution getting from the measurement the actual values can be achieved with the help of the numerical model by sensitivity analysis and fine tuning.

During the blackthorn drying the moisture content measurement was performed in traditional way. From the final moisture content the dry matter content was calculated first, and after that followed the relative moisture content on dry basis for the whole period. Because of the constant environmental circumstances the in-room drying process was considered to be a negative exponential time function which seems to be correct. The model has a 0,995 correlation coefficient between the natural logarithm of the moisture content and the time.

The changes of the area, the outline and the average diameter of the blackthorn against the moisture content are graphed in the Fig. 3, and the pictures of the same product in different times can be seen in Fig. 4.

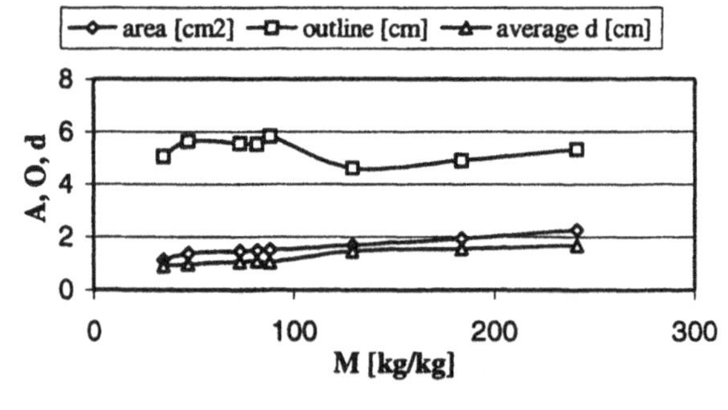

Figure 3. Outline, area and average diameter of blackthorn

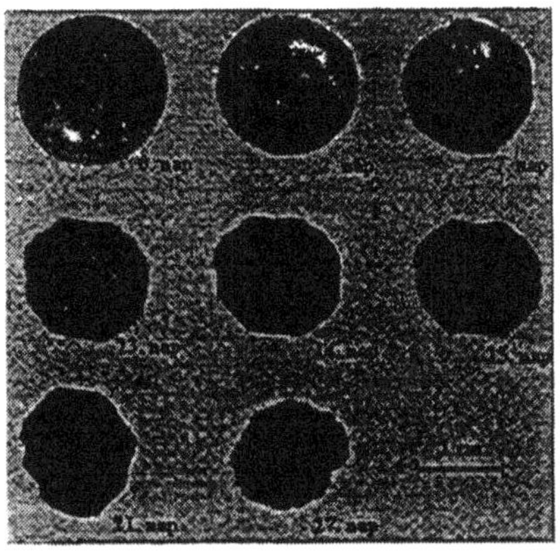

Fig. 4. Picture serial of a single blackthorn vs drying time in days

The change in the smoothness of the blackthorn is figured against the relative moisture content in Fig. 5 along with its typical picture shown in Fig. 6.

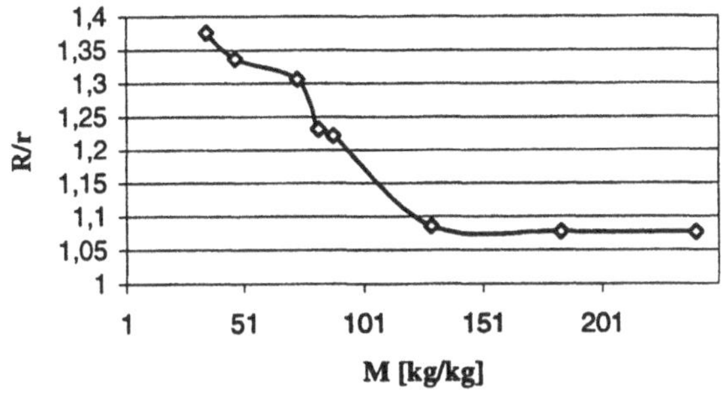

Fig. 5. The smoothness of blackthorn (R/r) vs moisture content

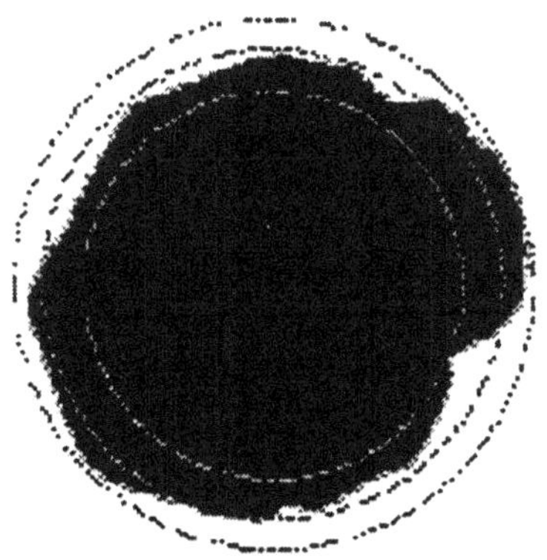

Fig. 6. Blackthorn with the outer [R] and inner[r] radius

Another quantity, the perpendicular area of the product is figured against the moisture content in Fig. 7.

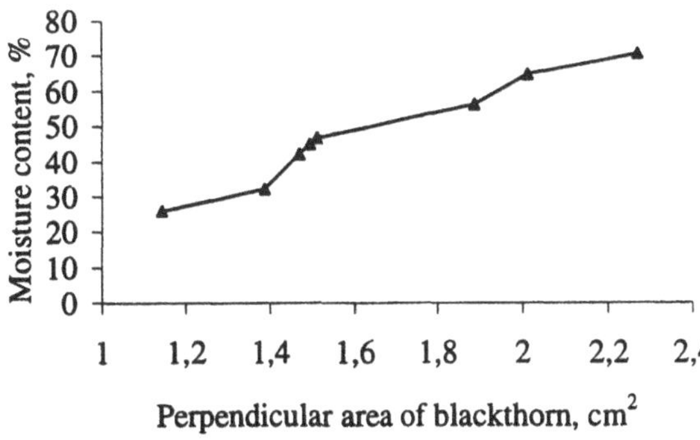

Fig. 7. The perpendicular area of blackthorn vs. moisture content

8. CONCLUSIONS

For the apple drying measurements the a physically based model was developed with a numerical solution. To validate the model, beside the traditional moisture content measurement the DNR technique could also be taken into the consideration. In such a way, it is possible to execute to determine the physical properties of the apple.

For the blackthorn measurement it can be concluded that the changes in the perpendicular area and in the average radius follows the changes in the moisture content by a linear function. It can be stated that the change of the outline does not follow the change of the moisture content because of the changes in the smoothness. During the analysis a mutually univocal connection was established between the moisture content of the product and the perpendicular area as well as the moisture content and the smoothness of the blackthorn. Based on these relations an optical moisture content measurement can be done.

Acknowledgement: This research was carried out with the support of the OTKA T 029300 and the MTA-TKI F-226/98 project. The blackthorn raw material used in the measurement was provided by the Bio Handwork Ltd.

NOTATION

C	- moisture content	kg/kg
C	- specific heat	J/kg K
D	- diffusion coefficient	m^2/s
D	- thickness of the skin	m
G	- mass flow rate	kg/m^2s
H	- relative humidity	
M	- moisture content on dry basis	kg/kg
R	- outer radius	m
R	- inner radius	m
$\mathfrak{R}$	- diffusion resistance	s/kg
t	- time variable	s
T	- air temperature	K

U	- heat transfer coefficient	W/m^3K
V	- volume of the pulp	m^3
X	- space variable	m

Greek Symbols

ρ	– density	kg/m^3
Θ	- product temperature	K
τ	- time constant	1/s

Subscripts

a	air	p	product
b	bulk	t	total
w	water	e	equilibrium

REFERENCES

Do, D.S., Sagara,Y., Kudoh, K., Yokota, H. and Higuchi, T. (1997). Surface Area and Volume Measurements of a Broccoli with a Microslicer-Image data Processing System, *5th International Symposium on Fruit, Nut and Vegetable Production Engineering,* Davis, California, USA, Session 12.

Farkas, I., Seres I. and Balló, B. (1996). Construction of a Modular Solar Dryer, *Proceedings of the 10th Drying Symposium (IDS'96),* **Vol. A,** pp. 585-590.

Murakami, M., Himoto, J. and Itoh, K. (1994). Analysis of Apple Quality by Near Infrared Reflectance Spectroscopy, *Journal of the Faculty of Agriculture, Hokkaido University,* **Vol. 66,** Pt. 1.

Raghavan, G.S.V. and Venkatachalapathy, K. (1999). Shrinkage of Strawberry during Microwave drying, *Drying Technology,* **Vol. 17,** No. 10, pp. 2309-2321.

Seres, I., Farkas, I. and Font, L. (2000). Experiences with use of solar energy in fruit drying, *Proceedings of the 12th International Drying Symposium (IDS2000),* Noordwijkerhout, The Netherlands

Sun, Y., Pantelides, C.C. and Chalabi, Z.S. (1995). Mathematical modeling and Simulation of Near-ambient Grain drying, *Computers and Electronics in Agriculture,* **Vol. 13,** pp. 243-271.

Whitelock, D.P., Brusewitz, G.H. and Ghajar, A.J. (1999). Thermal/Physical Properties Affect Predicted Weight Loss of Fresh Peaches, *American Society of Agricultural Engineers,* **Vol. 42,** No. 2, pp. 1047-1053.

IMPROVED CLIMATE CONTROL FOR POTATO STORES BY USING FUZZY CONTROLLERS

Klaus Gottschalk[1], László Nagy [2], István Farkas[2]

[1] *Institut für Agrartechnik Bornim e. V. ATB, Max-Eyth-Allee 100, D-14469 Potsdam, Germany,*
[2] *Szent István University Gödöllö, Faculty of Mechanical Engineering, Dept. of Physics and Process Control, Páter K. u. 1., H-2103 Hungary*

Abstract: The main tasks for climate controllers in potato bulk stores are to keep the storage climate in a constant state for quality conservation. To reduce energy costs the internal climate of potato bulk stores is controlled with outdoor air. Since a conventionally designed controller is difficult to adapt to reach an appropiate climate control, a controller designed with fuzzy logic is found as advantageous for adapting control parameters to improve the control process. A test stand is built to ventilate and control two samples of potato bulks simultaneously under the same outdoor weather conditions. Two different climate controllers (e.g. conventional to fuzzy) are compared.

Keywords: Fuzzy control, Potato store

1. INTRODUCTION

The most important quality factors for potatoes are freshness (low mass/water loss), health (dry storage) and absences of germs (cool storage). Bulk stores for potatoes are mostly ventilated by outdoor air (fig. 1). To avoid energy costs, cooling and heating systems are sometimes used temporarily only, when needed and applicable. Some cooling systems are mobile. When not using such systems, the store climate is controlled by outdoor climate only. Harvesting time and storing time for potatoes is about in September to October. In this time mostly there are cool nights and some cool days for using fresh cool air to ventilate the potato bulk directly. The basic control ‚rules' are defined as:

- *if* cooling air is needed
- *and*
- *if* cool air is available
- *then*
- ventilate (open dampers and switch on fans)
- *else*
- do not ventilate (close dampers and switch off fans)

Under Mid-European climatic conditions, the following objectives are to be fulfilled ('rules' to control the climate):

- dry the potato surfaces if nessesary
- cool down the potatoes as fast as possible after storage
- keep potato bulk cool during the whole storage time at constant temperature
- avoid wet surface i.e. avoid water condensation on the potatoes to prevent rot
- warm up the potatoes as fast as possible after storage time
- minimize mass loss during the whole storage time

- minimize energy consumption during the whole storage time

To achieve these aims, a well adapted control algorithm is to be implemented in the control equipment. Conventionally, the climate controller controls the fan and the damper openings to get inlet air flow to the potato bulk. The controller respects the current outdoor climate, i.e. outdoor temperature and outdoor air humidity (if available in practical stores). It takes also into account the (average) potato temperature, the inlet air temperature, the inlet air humidity (if available in practical stores) and outlet air temperature and finally outlet air humidity (if available in practical stores). In practical stores, the inlet air may be mixed with recirculated air to obtain the most suitable inlet air condition.

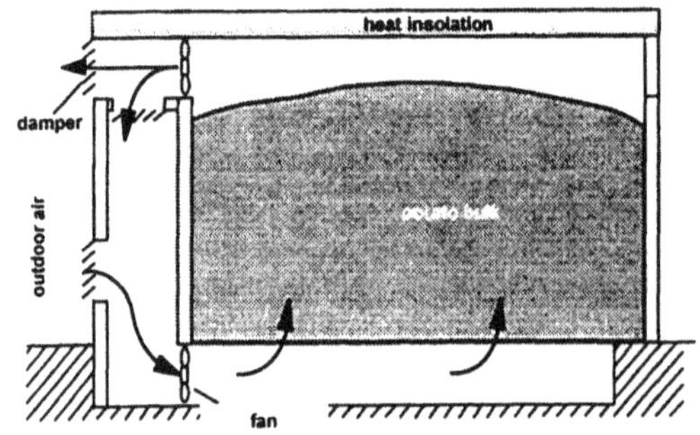

Fig. 1. Potato storage with fans and dampers.

2. CONVENTIONAL CONTROL

The climatic controller controls the damper positions (open - partially open - close positions) and the fan (on – off). The 'rules' to control the climate are forming a complex set of dependencies which are difficult to implement, to maintain and to test when using a conventional controller. A good comprehensible aid to manage these problems is to use state diagrams. The state diagram for fan control is shown in fig. 2, for example. More diagrams are defined for damper control, humidity control etc..

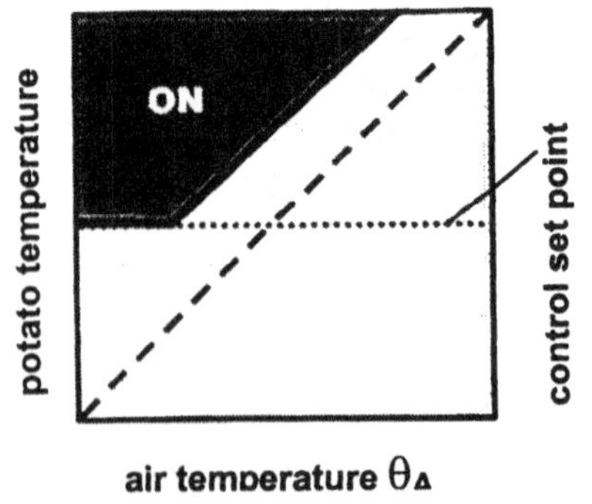

Fig. 2. State diagram for cooling control

On these state diagrams, the ‚intuitive‘ rules can be well defined and it can be seen how to act for cooling a storeroom (Maltry et al). The diagram area is seperated into different areas corresponding to different actions. A change of the ‚state‘ of the controller corresponds to a change of the state point (e.g. $\{\theta_P, \theta_A\}$) to another area. The border lines between the areas denote the ‚shift‘ of state. For example, if the outdoor temperature (θ_A) is below the potato temperature (θ_P), the state point lies in the area above the diagonal line. In this corresponding area the control fan action is defined to switch ON. When the potato temperature approches the line, it means that the temperature aprroches the storage control set point temperature. When the state point is close to the line or exceeds the line, the fan action is to switch OFF, etc.

For the conventional programming technique it is incomprehensible to implement the rules when the rules are complex (fig. 3).

```
if (Tair<Tpot) AND
NOT(Tinlet<Tdesire+Hyst) then begin

 if Tpot<5            then GOTO M1;
 if Tpot<Tair+Hyst    then GOTO M1;
 if FAN=ON            then GOTO M3;
 if Tpot<5.5          then GOTO M2;
 if Tpot<Tair+Hyst    then GOTO M2;
 FAN:=ON;
 if Tair < freezingPoint  then
     Damper=CLOSE
 else Damper=OPEN;

M3: if Tinlet<1.0     then
    Damper=CLOSE
 else
    if Tpot>Tinlet+7 then
    Damper=CloseOneStep
    else
    if Tpot<Tinlet+5 then
    Damper=OpenOneStep
 GOTO M_END;

M1: FAN=OFF;
M2: if Tpot>Tair      then
    if Tpot>5.5       then
    begin
      DAMPER=OPEN; GOTO M_END
    end;

M4: if Tpot<Tair-1    then
    DAMPER=CLOSE
end;
M END:
```

Fig. 3. Conventional fan control algorithm for cooling

3. FUZZY CONTROL

Regarding these circumstances gave the idea to use a fuzzy control algorithm to profit from the ability to implement the 'rules' directly into the rulebase of a fuzzy controller. A storekeeper may define a rule like this, for example: ‚IF PotatoTemperature=high AND AirTemperature=low THEN Fan=ON‘, and so on. A fuzzy controller comes therefore close to the intuitive manner of defining rules and measures.

The input values of the fuzzy controller are

- temperature difference between potatoes and inlet air, i.e. θ_{pot}-θ_{air} (with linguistic variable denoted DT)

- potato temperature (average value or from a sensor placed in approx. 1 m depth from bulk surface) θ_{pot} (ling. var. TP)
- outdoor air humidity, if applied
- inlet air humidity; probably mixed with outlet air, if applied.

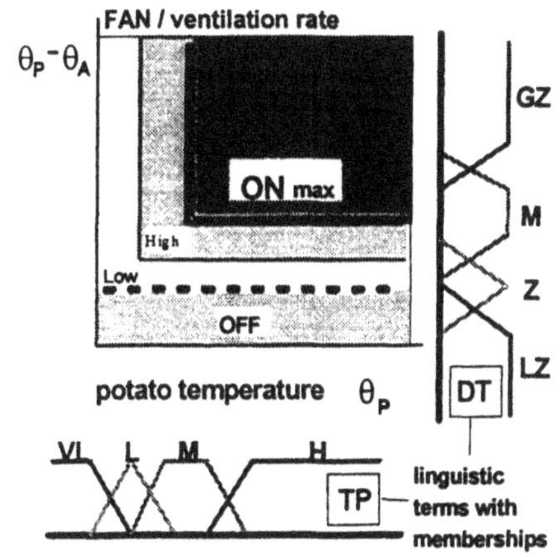

Fig. 4. Linguistic terms, membership functions and transfer response of the air rate fuzzy control

The output value (control value) is controlling the ventilation rate (continuous resp. discrete, i.e. 'ON/OFF'). The ventilation rate is a result from the answer of the fuzzy-controller respecting the input values and the 'climatization rules'. The respond function of the fuzzy-control algorithm for the ventilation rate, dependend on the temperature difference and the potato temperature, reflects the rules (fig. 4). The variables for the input values, e.g. of the temperature differences and the potato temperature are defined as linguistic variables DT resp. TP. For each of this linguistic variable, four fuzzy sets are defined. The definition of these four fuzzy sets is shown here as an example. The state point to switch the fan is about ±0 K for the temperature difference and about 4 °C for the potato temperature (the optimal storage temperature). If the fan is controlled by a frequency converter, the ventilation rate may well adapted to reach a ‚smooth' cooling process. This means for example that the ventilation rate should be reduced, when the potato temperature reaches its optimal storage temperature resp. when the difference temperature is about 0 K. To reduce the mass loss, additional rules are implemented to make use of a high inlet air humidity (resp. outdoor air humidity, if applied). In this case a higher ventilation rate is allowed when the air is of high humidity. When the inlet air is too dry, then the drying effect on the lower layers of the bulk must be reduced to avoid too much mass loss.

4. TEST STAND

A test stand is constructed to demonstrate climate control, fig. 5. The equipment reflects a real potato bulk storage facility. The potato samples are stored into two seperate sections which can be independently controlled. The two tests run concurrently with almost the same ambient conditions. With this configuration comparisons can be made between effects on different samples, or different control strategies, or different control algorithms, etc.

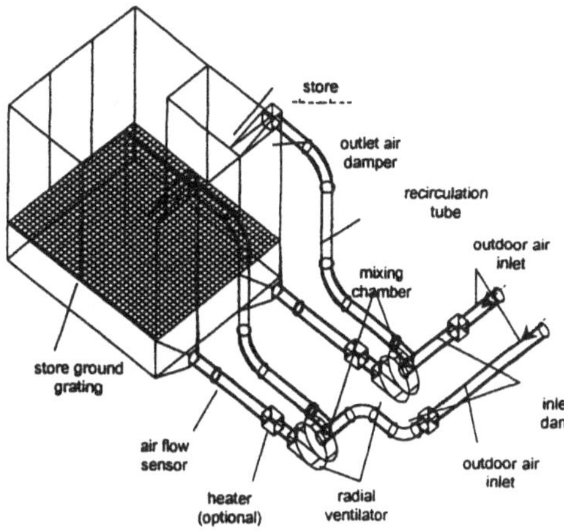

Fig. 5. Test stand with two sections

Sensors are installed to measure temperatures and relative humidity in the inlet air and the outlet air duct separately for each section. Up to 30 temperature sensors may be placed into the boxes to acquire bulk temperatures in several places within the bulk. At the outlet ducts and before the mixing chambers dampers are installed to control air flow mixing and recirculation. The fan motor revolutions can be remote controlled by frequency converters to control air rate. An industrial PC is controlling the equipment and logging the data automatically.

5. EXPERIMENTS

Test runs are made with several potato samples. In each section one box was placed which contained approx. 190 kg potatoes in 19 bags of 10 kg each. The box dimensions are: height=1.4 m and squared ground area=0.218 m². The bags where weighted before and after each test to obtain mass loss for the test run period. Five cooling down controlled processes were run (test #1 to #5), see table 1.

The first test (test #1) was run with conventional control on both sections (1) and (2), the other tests (test #2 – test #5) with fuzzy control on section (1), the other section (2) conventional controlled. Section (1) was always run with lower air rate (air velocity) than section (2). This shows lower temperature decrease rates (°C/hrs) for section (1) than for section (2). Test #2 and test #3 were run with fuzzy control on section (1) without optimization. Test #4 and test #5 were run with fuzzy control with optimizations made as discussed by Gottschalk (1997).

Table 1 Experiment results comparisons

TEST #	1	2	3	4	5
Fuzzy/Conv.	C/C	F/C	F/C	F/C	F/C
Runtime [hrs]	14.0	12.6	18.0	23.0	18.5
Fan run time [min] (1)	300	585	1095	1205	730
Fan run time [min] (2)	300	530	905	1120	550
Fan load [%] (1)	50.0	39.8	53.1	23.2	20.0
Fan load [%] (2)	51.7	33.0	49.9	15.0	15.0
Start temp [°C] (1)	8.50	10.40	14.20	12.30	11.50
Start temp [°C] (2)	10.50	10.30	14.20	12.30	11.50
end temp [°C] (1)	6.50	5.40	9.10	7.50	9.50
end temp [°C] (2)	7.20	4.50	7.70	8.20	10.00
Diff Temp [°C] (1)	2.00	5.00	5.10	4.80	2.00
Diff Temp [°C] (2)	3.30	5.80	6.50	4.10	1.50
temp rate [°C/h] (1)	0.14	0.40	0.28	0.21	0.11
temp rate [°C/h] (2)	0.24	0.46	0.36	0.18	0.08
mass loss [%] (1)	0.33	0.15	0.46	0.63	0.63
mass loss [%] (2)	0.32	0.16	0.47	0.64	0.64
loss per hr [%/h] (1)	0.024	0.012	0.03	0.03	0.034
loss per hr [%/h] (2)	0.023	0.013	0.03	0.03	0.034
loss control bag	0.37	0.16	0.44	1.10	1.10
air velocity [m/s] (1)	0.16	0.12	0.18	0.04	0.03
air velocity [m/s] (2)	0.37	0.25	0.36	0.14	0.14
fan power [W] (1)	13.56	7.85	11.08	1.03	0.50
fan power [W] (2)	52.35	27.59	35.58	7.14	7.14
fan energy [kWh] (1)	0.57	0.51	0.107	0.005	0.001
fan energy [kWh] (2)	2.25	1.34	0.268	0.020	0.010

In these tests, also the air velocities were decreased, but kept even lower for the fuzzy system than for the conventional system. In these tests (test #4 and test #5) the temperature drecrease rates (°C/hrs) were higher on the ‚optimized' fuzzy system compared to the conventional system. This phenomena was reverse to the other tests (#1 to #3). Also, on these tests, the energy consumptions for fan motors were significantly lower on the fuzzy system, due to the lower average air velocities through the bulks. The reason is the significant dependency of the air flow pressure drop when flowing through porous bulk material (distributed resistance). Nevertheless, the temperature decrease rates for the fuzzy control system are high enough to cool the bulk fast. During the experiments, the potoes startet to germ, which caused a higher mass loss during test #3 to test #5.

6. CONCLUSION

When cooling potato bulks with outdoor air it is difficult to maintain exact conditions in two seperate sections for comparison experiments. Meanwhile the mass losses for all experimental runs had no significant differences (even compared to a non-ventilated control bag), the effects on the cooling rate and the air velocity, resp. the fan motor energy consumptions show significant differences. The air rate is an important factor for energy saving optimization procedures. Also, the air rate influences the mass loss and the temperature decrease rate. Lowering the air rate and consequently the energy cost is possible by using a frequency converter for fun motor revolution remote and fuzzy control. Due to the short run times, compared to real long storing time, significant dependencies of the mass loss on some optimization procedures could not be shown. The reason for this is that mass loss optimization procedures in simulations showed that a mass loss may be reduced by approx. 1% to 1.5% during a long time storage period over 6 months. This result can not be verified by experiments for a short test run period on small potato samples over 24 hrs, for example.

Acknoledgements: This bilateral project is sponsored by the Hungarian Ministry (No. TET D-17/1998) and the German Ministry BML Proj. 03/98.

REFERENCES

Gottschalk, K. (1997). Adaptive Control to Optimize the Climate for Potato Storehouses. In: *IFAC Mathematical and Control Applications in Agriculture and Horticulture*, pp. 261-265. Hannover.

Maltry, W., Gottschalk, K. (1993). Thermisches Verhalten von Kartoffeln im belüfteten Lager. *Landtechnik* **7/93**, pp. 373-376.

AN ALGORITHM OF CHARACTERISTIC EXPANSION FOR FUZZY REASONING BASED ON TRIPLE I METHOD

Xiong Fanlun Li Shaowen Wang Rujing

Institute of Intelligent Machines, Chinese Academy of Sciences, Hefei 230031
E-mail: flxiong@163.net shwli@mail.hf.ah.cn

Abstract: Facing the defect in CRI method in logical semantics and the problem in its application to longer reasoning chaining in expert system, this paper, based on triple I method in literature (Wang, 1999a), proposes a characteristic expansion algorithm of fuzzy reasoning. The algorithm is then used in the fuzzy reasoning modes of multi-dimension and multi-implication, from which the models' characteristic expansion algorithms are deducted. As a result, a new way-out is found in the application of fuzzy reasoning in the intelligent systems. *Copyright© 2001 IFAC*

Key words: fuzzy reasoning, triple I method, characteristic expansion algorithm

1. INTRODUCTION

Since the approximate reasoning problem and its CRI (Compositional Rule of Inference) method were put forward in the literature (Zadeh, 1973), the study of fuzzy reasoning method has been developing very fast. Many methods in the definition of fuzzy relation and compositional operation (Mizumoto and Zimmerman, 1982), and other methods for the improvement of CRI have been proposed. And fuzzy reasoning based on CRI-method has been constantly applied in various ideas. As a result, as many as 100 methods are found in today's literature (Dubois, *et al.*, 1991; Wu, 1994). They have become one of the important branches in fuzzy system theory.

However, seen from logical semantics, CRI method is a single implication algorithm, which actually uses fuzzy reasoning for only once. But implication algorithm, which should be further used, is replaced by compositional operation simply. It lacks evidence. And it is why there is difficulty in bringing CRI into logical framework (Elkan, 1994). Facing the defect in CRI method, a new concept of fuzzy reasoning based on triple implication (Triple I) was proposed in the literature (Wang, 1999a), and this breakthrough is an important contribution to fuzzy reasoning in theory. But it is difficult to be applied in the intelligent systems, especially in the longer reasoning chaining of expert systems. Therefor, this paper proposes a new algorithm of the characteristic expansion based on triple I method, so as to provide the possibility of fuzzy reasoning application in intelligent systems.

Supported by the National 863 Project of China (No. 863-306-ZD05-03-4)

2. THE BASIC THOUGHT OF TRIPLE I REASONING AND THE ALGORITHM OF CHARACTERISTIC EXPANSION

If $\underset{\sim}{A}, \underset{\sim}{A}^* \in F(U), \underset{\sim}{B}, \underset{\sim}{B}^* \in F(V)$ are given, then FMP (Fuzzy Modus Ponens) reasoning can be expressed as:

$$\begin{array}{lll} \textit{Known} & \underset{\sim}{A} \to \underset{\sim}{B} & \\ \textit{and Given} & \underset{\sim}{A}^{*} & . \quad (1) \\ \hline \textit{Searching} & \underset{\sim}{B}^{*} & \end{array}$$

FMT (Fuzzy Modus Tollens) reasoning can be expressed as:

$$\begin{array}{lll} \textit{Known} & \underset{\sim}{A} \to \underset{\sim}{B} & \\ \textit{and Given} & \underset{\sim}{B}^{*} & . \quad (2) \\ \hline \textit{Searching} & \underset{\sim}{A}^{*} & \end{array}$$

The basic thought based on the fuzzy reasoning algorithm of triple I method is that optimal $\underset{\sim}{B}^{*} \in F(V)$ or $\underset{\sim}{A}^{*} \in F(U)$ is searched in order to meet the largest sustentation degree of $\underset{\sim}{A} \to \underset{\sim}{B}$ for $\underset{\sim}{A}^{*} \to \underset{\sim}{B}^{*}$ when $\underset{\sim}{A} \in F(U)$, $\underset{\sim}{B} \in F(V)$ and $\underset{\sim}{A}^{*} \in F(U)$ (or $\underset{\sim}{B}^{*} \in F(V)$) are known. The general form of this algorithm can be expressed by the optimum problem:

For $\omega \in [0,1]$, when $\underset{\sim}{A}$, $\underset{\sim}{B}$ and $\underset{\sim}{A}^{*}$ (or $\underset{\sim}{B}^{*}$) are known, optimal $\underset{\sim}{B}^{*}$ (or $\underset{\sim}{A}^{*}$) is searched so as to meet

$$(\underset{\sim}{A}(u) \to \underset{\sim}{B}(v)) \to (\underset{\sim}{A}^{*}(u) \to \underset{\sim}{B}^{*}(v)) \geq \omega \quad (3)$$

under $(u,v) \in U \times V$. Here ω is known as sustentation degree for $\underset{\sim}{A}^{*} \to \underset{\sim}{B}^{*}$ on $\underset{\sim}{A} \to \underset{\sim}{B}$.

The character of five kinds of implication operators was analyzed in the literature (Wang, 1999a, b) and the ω-triple I principles were reached as follows.

ω-triple I principle (FMP) If U,V are not empty set and $\underset{\sim}{A}, \underset{\sim}{A}^{*} \in F(U)$, $\underset{\sim}{B} \in F(V)$ are known, then $\underset{\sim}{B}^{*}$ in Form (1) is minimal fuzzy set which makes Form (3) come into existence in $F(V)$. $\underset{\sim}{B}^{*}$ is known as the ω-solution of Form (1).

ω-triple I principle (FMT) If U,V are not empty set and $\underset{\sim}{A} \in F(U)$, $\underset{\sim}{B}, \underset{\sim}{B}^{*} \in F(V)$ are known, then $\underset{\sim}{A}^{*}$ in Form (2) is maximal fuzzy set which makes Form (3) come into existence in $F(U)$. $\underset{\sim}{A}^{*}$ is called as the ω-solution of Form (2).

It is discovered that the Mamdani's implication operator, $\underset{\sim}{R}_c$, is not only simple but also reductive. So, by using $\underset{\sim}{R}_c$, a generalizing discussion will be made for the algorithm of characteristic expansion of triple I reasoning. Here only the FMP mode is discussed, and the discussion about FMT model is similar to FMP's. But it should be pointed out that, because $\underset{\sim}{R}_c(a,b) = a \wedge b, \ (a,b \in [0,1])$ is not decreasing monotonously with a and b, and it is different from the implication operators in the literature (Wang, 1999a, b), the $\underset{\sim}{R}_c$-solution in ω-triple I principle (FMT), $\underset{\sim}{A}^{*} \in F(U)$, is the minimal fuzzy set which makes Form (3) come into existence in $F(U)$, not the maximal fuzzy set.

Theorem 1 If $\underset{\sim}{A}, \underset{\sim}{A}^{*} \in F(U)$ and $\underset{\sim}{B} \in F(V)$ are given, then, for the FMP in Form (1), the characteristic expansion algorithm of $\underset{\sim}{B}^{*}$, which meets with Form (3), is

$$\underset{\sim}{B}^{*} = (\omega \wedge \alpha) \cdot \underset{\sim}{B}. \quad (4)$$

In Form (4), the $\alpha = \sup_{u \in U}[\underset{\sim}{A}^{*}(u) \wedge \underset{\sim}{A}(u)]$ is known as the characteristic coefficient of $\underset{\sim}{A}^{*}$.

Proof: If $\underset{\sim}{R}_c$ is known as implication operator, then Form (3) becomes:

$$\begin{aligned} &(\underset{\sim}{A}(u) \xrightarrow[c]{} \underset{\sim}{B}(v)) \xrightarrow[c]{} (\underset{\sim}{A}^{*}(u) \xrightarrow[c]{} \underset{\sim}{B}^{*}(v)) \\ &= \underset{\sim}{R}_c(\underset{\sim}{A}(u), \underset{\sim}{B}(v)) \xrightarrow[c]{} \underset{\sim}{R}_c(\underset{\sim}{A}^{*}(u), \underset{\sim}{B}^{*}(v)) \\ &= \underset{\sim}{R}_c(\underset{\sim}{A}(u), \underset{\sim}{B}(v)) \wedge \underset{\sim}{R}_c(\underset{\sim}{A}^{*}(u), \underset{\sim}{B}^{*}(v)) \\ &= \underset{\sim}{R}_c(\underset{\sim}{A}(u), \underset{\sim}{B}(v)) \wedge [\underset{\sim}{A}^{*}(u) \wedge \underset{\sim}{B}^{*}(v)] \\ &= [\underset{\sim}{A}^{*}(u) \wedge \underset{\sim}{R}_c(\underset{\sim}{A}(u), \underset{\sim}{B}(v))] \wedge \underset{\sim}{B}^{*}(v) \geq \omega. \quad (5) \end{aligned}$$

From Form (5), the following form can be deduced:

$$\underset{\sim}{B}^{*}(v) \geq [\underset{\sim}{A}^{*}(u) \wedge \underset{\sim}{R}_c(\underset{\sim}{A}(u), \underset{\sim}{B}(v))] \wedge \underset{\sim}{B}^{*}(v) \geq$$

$$\omega \geq \omega \wedge \sup_{u \in U}[\underset{\sim}{A}^{*}(u) \wedge \underset{\sim}{R}_c(\underset{\sim}{A}(u), \underset{\sim}{B}(v))]$$

Thus,

$$\underset{\sim}{B}^{*}(v) \geq \omega \wedge \sup_{u \in U}[\underset{\sim}{A}^{*}(u) \wedge \underset{\sim}{R}_c(\underset{\sim}{A}(u), \underset{\sim}{B}(v))].$$

According to ω-triple I principle (FMP), here $\underset{\sim}{B}^{*}$

$\in F\ (V)$ should be a minimal fuzzy set. So $\underset{\sim}{B}^*(v)$ adopts:

$$\underset{\sim}{B}^*(v) = \omega \wedge \sup_{u \in U}[\underset{\sim}{A}^*(u) \wedge \underset{\sim}{R}_c(\underset{\sim}{A}(u), \underset{\sim}{B}(v))], v \in V \tag{6}$$

namely, $\underset{\sim}{B}^* = \omega \cdot (\underset{\sim}{A}^* \circ \underset{\sim}{R}_c) = \omega \cdot \underset{\sim}{A}^* \circ \underset{\sim}{R}_c$.

Form (6) can be further analyzed, and then its characteristic expansion algorithm is reached:

$$\begin{aligned}\underset{\sim}{B}^*(v) &= \omega \wedge \sup_{u \in U}[\underset{\sim}{A}^*(u) \wedge \underset{\sim}{A}(u) \wedge \underset{\sim}{B}(v)] \\ &= \omega \wedge \sup_{u \in U}[\underset{\sim}{A}^*(u) \wedge \underset{\sim}{A}(u)] \wedge \underset{\sim}{B}(v) \\ &= \omega \wedge \alpha \wedge \underset{\sim}{B}(v) = (\omega \wedge \alpha) \wedge \underset{\sim}{B}(v).\end{aligned} \tag{7}$$

Namely, $\underset{\sim}{B}^* = (\omega \wedge \alpha) \cdot \underset{\sim}{B}$.

Here the $\alpha = \sup_{u \in U}[\underset{\sim}{A}^*(u) \wedge \underset{\sim}{A}(u)]$ is the characteristic coefficient of $\underset{\sim}{A}^*$. Hence, the Theorem 1 is proved.

3. THE CHARACTERISTIC EXPANSION ALGORITHM IN MULTI-IMPLICATION AND MULTI-DIMENSION BASED ON TRIPLE I FUZZY REASONING

The fuzzy reasoning in multi-implication and multi-dimension, which includes several kinds of fuzzy reasoning modes such as multi-implication, multi-dimension, and multi-implication and multi-dimension, is the extension of basic fuzzy reasoning. Next, their characteristic expansion algorithms are discussed respectively.

3.1 The characteristic expansion algorithm in the multi-dimension based on triple I fuzzy reasoning

Its FMP reasoning mode is

$$\begin{array}{ll} \textit{Known} & \underset{\sim}{A}_1\ and\ \underset{\sim}{A}_2\ and \cdots and\ \underset{\sim}{A}_m \to \underset{\sim}{B} \\ \textit{and Given} & \underset{\sim}{A}_1^*\ and\ \underset{\sim}{A}_2^*\ and \cdots and\ \underset{\sim}{A}_m^* \\ \hline \textit{Searching} & \underset{\sim}{B}^* \end{array} \tag{8}$$

If $\underset{\sim}{A}_j, \underset{\sim}{A}_j^* \in F\ (U_j)$, $\underset{\sim}{B} \in F\ (V)$ $(j = 1,2, \cdots, m)$ are given, and

$$\underset{\sim}{A}_1\ and\ \underset{\sim}{A}_2\ and \cdots and\ \underset{\sim}{A}_m = \underset{\sim}{A}_1 \cap \underset{\sim}{A}_2 \cap \cdots \cap \underset{\sim}{A}_m = \underset{\sim}{A}_1 \times \underset{\sim}{A}_2 \times \cdots \times \underset{\sim}{A}_m,$$

$$\underset{\sim}{A}_1^*\ and\ \underset{\sim}{A}_2^*\ and \cdots and\ \underset{\sim}{A}_m^* = \underset{\sim}{A}_1^* \cap \underset{\sim}{A}_2^* \cap \cdots \cap \underset{\sim}{A}_m^* = \underset{\sim}{A}_1^* \times \underset{\sim}{A}_2^* \times \cdots \times \underset{\sim}{A}_m^*$$

are assumed, then fuzzy reasoning based on triple I can be expressed as:

$$[\bigwedge_{j=1}^{m} \underset{\sim}{A}_j(u_j) \to \underset{\sim}{B}(v)] \to [\bigwedge_{j=1}^{m} \underset{\sim}{A}_j^*(u_j) \to \underset{\sim}{B}^*(v)] \geq \omega \tag{9}$$

comes into existence for $u_j \in U_j (j = 1,2, \cdots, m)$ and $v \in V$.

Theorem 2 If $\underset{\sim}{A}_j, \underset{\sim}{A}_j^* \in F\ (U_j)$ and $\underset{\sim}{B} \in F\ (V)$ $(j = 1,2, \cdots, m)$ are given, then, for the FMP in Form (8), the characteristic expansion algorithm of $\underset{\sim}{B}^*$, which meets with Form (9), is

$$\underset{\sim}{B}^* = (\omega \wedge \bigwedge_{j=1}^{m} \alpha_j) \cdot \underset{\sim}{B}. \tag{10}$$

In Form (10), the $\alpha_j = \sup_{u_j \in U_j}[\underset{\sim}{A}_j^*(u_j) \wedge \underset{\sim}{A}_j(u_j)]$ is known as the characteristic coefficient of $\underset{\sim}{A}_j^*$.

Proof: If $\underset{\sim}{R}_c$ known as implication operator, and

$$\underset{\sim}{A}(u_1, u_2, \cdots, u_m) = \bigwedge_{j=1}^{m} \underset{\sim}{A}_j(u_j),$$

$$\underset{\sim}{A}^*(u_1, u_2, \cdots, u_m) = \bigwedge_{j=1}^{m} \underset{\sim}{A}_j^*(u_j)$$

assumed, then Form (9) can be converted into

$$[\underset{\sim}{A}(u_1, u_2, \cdots, u_m) \xrightarrow[c]{} \underset{\sim}{B}(v)] \xrightarrow[c]{} [\underset{\sim}{A}^*(u_1, u_2, \cdots, u_m) \xrightarrow[c]{} \underset{\sim}{B}^*(v)] \geq \omega. \tag{11}$$

Form (11) and Form (3) are alike, if the former is compared with latter. So the characteristic expansion algorithm based on Form (11) identifies with Form (7), namely:

$$\underset{\sim}{B}^*(v) = (\omega \wedge \alpha) \wedge \underset{\sim}{B}(v), \tag{12}$$

that is to say, $\underset{\sim}{B}^* = (\omega \wedge \alpha) \cdot \underset{\sim}{B}$. But here, the characteristic coefficient is

$$\alpha = \sup_{(u_1,u_2,\cdots,u_m)\in \times_{j=1}^{m} U_j} [\underset{\sim}{A}^*(u_1,u_2,\cdots,u_m) \wedge \underset{\sim}{A}(u_1,u_2,\cdots,u_m)]$$

$$= \sup_{(u_1,u_2,\cdots,u_m)\in \times_{j=1}^{m} U_j} [\wedge_{j=1}^{m} \underset{\sim}{A}_j^*(u_j) \wedge \wedge_{j=1}^{m} \underset{\sim}{A}_j(u_j)]$$

$$= \sup_{(u_1,u_2,\cdots,u_m)\in \times_{j=1}^{m} U_j} \{\wedge_{j=1}^{m} [\underset{\sim}{A}_j^*(u_j) \wedge \underset{\sim}{A}_j(u_j)]\}$$

$$= \wedge_{j=1}^{m} \{\sup_{u_j \in U_j} [\underset{\sim}{A}_j^*(u_j) \wedge \underset{\sim}{A}_j(u_j)]\} = \wedge_{j=1}^{m} \alpha_j, \quad (13)$$

In Form (13), the $\alpha_j = \sup_{u_j\in U_j}[\underset{\sim}{A}_j^*(u_j) \wedge \underset{\sim}{A}_j(u_j)]$ is known as the characteristic coefficient of $\underset{\sim}{A}_j^*$.

If Form (13) is put into Form (12), then the characteristic expansion algorithm in the multi-dimension based on triple I fuzzy reasoning will be confirmed:

$$\underset{\sim}{B}^*(v) = (\omega \wedge \wedge_{j=1}^{m} \alpha_j) \wedge \underset{\sim}{B}(v). \quad (14)$$

Namely, $\underset{\sim}{B}^* = (\omega \wedge \wedge_{j=1}^{m} \alpha_j) \cdot \underset{\sim}{B}$. Hence, the Theorem 2 is proved.

3.2 The characteristic expansion algorithm in the multi-implication based on triple I fuzzy reasoning

Its FMP reasoning mode is:

$$\begin{array}{ll} \textit{Known} & \underset{\sim}{A}_1 \to \underset{\sim}{B}_1 \\ & \cdots\cdots \\ & \underset{\sim}{A}_n \to \underset{\sim}{B}_n \\ \textit{and Given} & \underset{\sim}{A}^* \\ \hline \textit{Searching} & \underset{\sim}{B}^* \end{array} \quad . \quad (15)$$

According to the difference of aggregate forms, there are four kinds of methods for the solution of this mode (Li and Xiong, 2000). Here the method of FITA-And (First Infer Then aggregate-And) is selected, and then the theorem will be:

Theorem 3 If $\underset{\sim}{A}_i, \underset{\sim}{A}^* \in F\ (U)$ and $\underset{\sim}{B}_i \in F\ (V)$ $(i = 1,2,\cdots,n)$ are given, then, for the FMP in Form (15), the characteristic expansion algorithm of $\underset{\sim}{B}^*$, which is based on triple I and FITA-And method, is

$$\underset{\sim}{B}^* = \bigcup_{i=1}^{n} (\omega_i \wedge \alpha_i) \cdot \underset{\sim}{B}_i. \quad (16)$$

Here ω_i is the sustentation degree for $\underset{\sim}{A}^* \to \underset{\sim}{B}_i^*$ on $\underset{\sim}{A}_i \to \underset{\sim}{B}_i$; $\alpha_i = \sup_{u\in U}[\underset{\sim}{A}^*(u) \wedge \underset{\sim}{A}_i(u)]$ is known as the characteristic coefficient of $\underset{\sim}{A}^*$.

Proof: For $\underset{\sim}{A}_i, \underset{\sim}{A}^* \in F\ (U)$ and $\underset{\sim}{B}_i \in F\ (V)$ $(i = 1,2,\cdots,n)$, the solution of FITA-And method can be expressed as:

$$\underset{\sim}{B}^* = \bigcup_{i=1}^{n} \underset{\sim}{B}_i^*. \quad (17)$$

Here $\underset{\sim}{B}_i^* \in F\ (V)$ is the solution of

$$\begin{array}{ll} \textit{Known} & \underset{\sim}{A}_i \to \underset{\sim}{B}_i \\ \textit{and Given} & \underset{\sim}{A}^* \\ \hline \textit{Searching} & \underset{\sim}{B}_i^* \end{array} \quad . \quad (18)$$

Compared with Form (1), Form (18) based on triple I reasoning can be expressed as:

$$(\underset{\sim}{A}_i(u) \to \underset{\sim}{B}_i(v)) \to (\underset{\sim}{A}^*(u) \to \underset{\sim}{B}_i^*(v)) \ge \omega_i \quad . \quad (19)$$

Then the characteristic expansion algorithm should be:

$$\underset{\sim}{B}_i^*(v) = (\omega_i \wedge \alpha_i) \wedge \underset{\sim}{B}_i(v),$$

namely,

$$\underset{\sim}{B}_i^* = (\omega_i \wedge \alpha_i) \cdot \underset{\sim}{B}_i. \quad (20)$$

Here $\alpha_i = \sup_{u\in U}[\underset{\sim}{A}^*(u) \wedge \underset{\sim}{A}_i(u)]$.

If Form (20) put into Form (17), the characteristic expansion algorithm in the multi-implication based on triple I fuzzy reasoning can be:

$$\underset{\sim}{B}^* = \bigcup_{i=1}^{n} (\omega_i \wedge \alpha_i) \cdot \underset{\sim}{B}_i,$$

namely, $\underset{\sim}{B}^*(v) = \overset{n}{\underset{i=1}{\vee}}[(\omega_i \wedge \alpha_i) \wedge \underset{\sim}{B}_i(v)]$.

3.3 The characteristic expansion algorithm in multi-implication and multi-dimension based on triple I fuzzy reasoning

Its FMP reasoning mode is:

$$\begin{array}{ll} \textit{Known} & \underset{\sim}{A}_{11}\, and \underset{\sim}{A}_{12}\, and \cdots and \underset{\sim}{A}_{1m} \rightarrow \underset{\sim}{B}_1 \\ & \cdots\cdots \\ & \underset{\sim}{A}_{n1}\, and \underset{\sim}{A}_{n2}\, and \cdots and \underset{\sim}{A}_{nm} \rightarrow \underset{\sim}{B}_n \\ \textit{and Given} & \underset{\sim}{A}_1^*\, and \underset{\sim}{A}_2^*\, and \cdots and \underset{\sim}{A}_m^* \\ \hline \textit{Searching} & \underset{\sim}{B}^* \end{array} \quad (21)$$

Given $\underset{\sim}{A}_{ij}, \underset{\sim}{A}_j^* \in F\ (U_j)$, $\underset{\sim}{B}_i \in F\ (V)\ (i = 1,2, \cdots, n; j = 1,2,\cdots,m)$ and

1) $\underset{\sim}{A}_{i1}$ *and* $\underset{\sim}{A}_{i2}$ *and* $\cdots$ *and* $\underset{\sim}{A}_{im}$

$= \underset{\sim}{A}_{i1} \cap \underset{\sim}{A}_{i2} \cap \cdots \cap \underset{\sim}{A}_{im}$

$= \underset{\sim}{A}_{i1} \times \underset{\sim}{A}_{i2} \times \cdots \times \underset{\sim}{A}_{im} \underline{\underline{\Delta}} \underset{\sim}{A}_i$, namely,

$\underset{\sim}{A}_i(u_1,u_2,\cdots,u_m) = \overset{m}{\underset{j=1}{\wedge}} \underset{\sim}{A}_{ij}(u_j), (i = 1,2,\cdots,n)$;

2) $\underset{\sim}{A}_1^*$ *and* $\underset{\sim}{A}_2^*$ *and* $\cdots$ *and* $\underset{\sim}{A}_m^*$

$= \underset{\sim}{A}_1^* \cap \underset{\sim}{A}_2^* \cap \cdots \cap \underset{\sim}{A}_m^*$

$= \underset{\sim}{A}_1^* \times \underset{\sim}{A}_2^* \times \cdots \times \underset{\sim}{A}_m^* \underline{\underline{\Delta}} \underset{\sim}{A}^*$, namely,

$\underset{\sim}{A}^*(u_1,u_2,\cdots,u_m) = \overset{m}{\underset{j=1}{\wedge}} \underset{\sim}{A}_j^*(u_j)$,

then Form (21) can be converted into Form (15). While the characteristic expansion algorithm of Form (15) is

$$\underset{\sim}{B}^* = \bigcup_{i=1}^{n} (\omega_i \wedge \alpha_i) \cdot \underset{\sim}{B}_i \,. \quad (22)$$

But here

$$\alpha_i = \sup_{(u_1,u_2,\cdots,u_m) \in \overset{m}{\underset{j=1}{\times}} U_j} [\underset{\sim}{A}^*(u_1,u_2,\cdots,u_m) \wedge \underset{\sim}{A}_i(u_1,u_2,\cdots,u_m)]$$

$$= \sup_{(u_1,u_2,\cdots,u_m) \in \overset{m}{\underset{j=1}{\times}} U_j} [\overset{m}{\underset{j=1}{\wedge}} \underset{\sim}{A}_j^*(u_j) \wedge \overset{m}{\underset{j=1}{\wedge}} \underset{\sim}{A}_{ij}(u_j)]$$

$$= \sup_{(u_1,u_2,\cdots,u_m) \in \overset{m}{\underset{j=1}{\times}} U_j} \{\overset{m}{\underset{j=1}{\wedge}} [\underset{\sim}{A}_j^*(u_j) \wedge \underset{\sim}{A}_{ij}(u_j)]\}$$

$$= \overset{m}{\underset{j=1}{\wedge}} \{\sup_{u_j \in U_j} [\underset{\sim}{A}_j^*(u_j) \wedge \underset{\sim}{A}_{ij}(u_j)]\}$$

$$= \overset{m}{\underset{j=1}{\wedge}} \alpha_{ij} \,. \quad (23)$$

And the characteristic coefficient of $\underset{\sim}{A}_j^*$ is

$$\alpha_{ij} = \sup_{u_j \in U_j} [\underset{\sim}{A}_j^*(u_j) \wedge \underset{\sim}{A}_{ij}(u_j)] .$$

If Form (23) put into Form (22), the characteristic expansion algorithm in multi-implication and multi-dimension based on triple I fuzzy reasoning can be:

$$\underset{\sim}{B}^* = \bigcup_{i=1}^{n} (\omega_i \wedge \overset{m}{\underset{j=1}{\wedge}} \alpha_{ij}) \cdot \underset{\sim}{B}_i \,,$$

namely, $\underset{\sim}{B}^*(v) = \overset{n}{\underset{i=1}{\vee}}[(\omega_i \wedge \overset{m}{\underset{j=1}{\wedge}} \alpha_{ij}) \wedge \underset{\sim}{B}_i(v)]$.

Hence:

Theorem 4 If $\underset{\sim}{A}_{ij}, \underset{\sim}{A}_j^* \in F\ (U_j)$ and $\underset{\sim}{B}_i \in F\ (V)$ $(i = 1,2,\cdots,n; j = 1,2,\cdots,m)$ are given, then, for the FMP in Form (21), the characteristic expansion algorithm of $\underset{\sim}{B}^*$ is

$$\underset{\sim}{B}^* = \bigcup_{i=1}^{n} (\omega_i \wedge \overset{m}{\underset{j=1}{\wedge}} \alpha_{ij}) \cdot \underset{\sim}{B}_i \,. \quad (24)$$

Here ω_i is the sustentation degree for $(\overset{m}{\underset{j=1}{\cap}} \underset{\sim}{A}_j^* \rightarrow \underset{\sim}{B}_i^*)$ on $(\overset{m}{\underset{j=1}{\cap}} \underset{\sim}{A}_{ij} \rightarrow \underset{\sim}{B}_i)$; $\alpha_{ij} = \sup_{u_j \in U_j}[\underset{\sim}{A}_j^*(u_j) \wedge \underset{\sim}{A}_{ij}(u_j)]$ is known as the characteristic coefficient of $\underset{\sim}{A}_j^*$.

4. CONCLUSION

In accordance with the defect in CRI method in logical semantics and the problem in its application to longer reasoning chaining in expert system, the fuzzy reasoning algorithm of characteristic expansion based on triple I method is proposed in this paper. The sets of this algorithm, which can be used in the fuzzy reasoning modes of multi-dimension and multi-implication, are also deducted. Based on the above research, a fuzzy reasoning model for imprecise spreading in agricultural expert systems has been established, which is dealt with in another paper.

REFERENCES

Dubois, D., H. Prade and J. Lang (1991). Fuzzy sets in approximate reasoning, part 1-2. *Fuzzy Sets and Systems,* **40**, 143-244.

Elkan, C. (1994). The paradoxical success of fuzzy logic. *IEEE Trans. Expert,* **9**, 3-8.

Li, S. W. and F. L. Xiong (2000). A study on the multi-rule fuzzy reasoning and its characteristic expansion algorithm. *J. of Biomathematics,* **15**(4), 1-8.

Mizumoto, M. and H. J. Zimmerman (1982). Comparison of fuzzy reasoning methods. *Fuzzy Sets and Systems,* **8**, 253-283.

Wang, G. J. (1999a). Triple I algorithm with totally inference rules of fuzzy reasoning. *Science in China, Series E,* **29**(1), 43-53.

Wang, G. J. (1999b). A new method for fuzzy reasoning. *Fuzzy Systems and Mathematics,* **13**(3), 1-12.

Wu, W. M. (1994). *Principles and Methods of Fuzzy Reasoning*. Guizhou Science and Technology Press, Guiyang.

Zadeh, L. A. (1973). Outline of a new approach to the analysis of complex systems and decision processes. *IEEE Trans. Systems, Man and Cybernetic,* **3**, 28-44.

APPLYING THE FUZZY MULTI-ATTRIBUTE DECISION MODEL IN PLANT BREEDING PROGRAMS

M. Calin, C. Leonte

University of Agricultural Sciences and Veterinary Medicine, Iasi, Romania

Decision making in environments that do not allow much algorithmic modelling is not easy. Biological sciences are such environments. More difficulties arise when some knowledge is expressed through linguistic terms instead of numeric values. Different approaches were developed for such decision situations. The paper refers to one of these methods, the Fuzzy Multi-Attribute Decision Model and suggests its utilisation as decision support in the selection phase of plant breeding programs. Representation and handling of linguistic terms is particularly focused. The results of a case study on a breeding program that involved a variety of pod bean are also presented. *Copyright © 2001 IFAC*

Keywords: Agriculture, Plants, Decision making, Decision support systems, Fuzzy modelling, Linguistic variables, Membership functions.

1. INTRODUCTION

The strategy of a plant breeding program is determined by the proposed goals and the biological characters of the involved species. Through artificial selection the *elite*, that is the plants or groups of plants that meet the requirements, are separated and used for continuing the breeding program. Selection is involved in any plant breeding program and its efficiency is conditioned by many factors. Among them, very important is the ability of the expert to identify the individuals that match the goals for the most of the considered characters. Usually, a large number of individuals are taken into consideration and the observed characters are also quite numerous, which brings the difficulty of processing a large amount of numerical data in order to make the selection. For this reason, the improvement of the selection methodology in a way that would enable its implementation within a software decision support tool could be very useful.

In the paper, the selection of the elite from a set of observed individuals is viewed as a decision problem with multiple objectives where the decision maker has to choose from a finite number of decision alternatives. An usual approach of such a problem is the Multi-Attribute Decision Model (MADM). Moreover, the decision maker often uses linguistic terms in the decision process. These linguistic terms must be properly formalised and, in this respect, the Fuzzy Sets Theory provides very useful tools. The result of the inclusion of Fuzzy Sets concepts within the MADM is a Fuzzy Multi-Attribute Decision Model (Zimmermann, 1996).

Since the selection criteria do not generally have the same level of importance, their evaluation is another important point of discussion. An efficient and quite easy way of performing this evaluation was suggested by Saaty (1980).

1.1 Fuzzy Terms

Like in other fields of activity, a plant breeding professional often uses *linguistic terms* instead of *crisp*, numerical values. He properly deals with qualifiers like "temperature is *low*", "concentration is *approximately* 5%", "beans' weight is *medium*", despite their imprecise appearance. If such terms are to be handled within a software tool, such as a *decision support system*, they must be formalised. Fuzzy Sets Theory and Fuzzy Logic provide mathematical support for this aim. Different branches of biological sciences are also known as fields of activity where linguistic terms are often manipulated and their representation through fuzzy sets can be done (Calin *et al.*, 1998).

A fuzzy set is a mapping $F: U \rightarrow [0, 1]$ where U is the *universe of discourse* and for any $x \in U$, $F(x)$ is the *membership degree* of x in F. A linguistic expression can be represented as a *fuzzy set* whose universe of discourse is defined by the referred element (temperature, concentration, weight, etc.). This is why the names *fuzzy term* and *linguistic term* are considered to be equivalent. The membership function $F(x)$ describes the degree in which, for any value in U, the linguistic statement may be considered to be true. Several *fuzzy terms* can be defined to cover the entire universe of discourse with non-zero values, that is a *fuzzy variable* (Reusch, 1996).

The definition of fuzzy terms and variables can also rely on other criteria like subjective belief, or prior experience. The case study included in the paper contains such situations.

1.2 The Fuzzy Multi-Attribute Decision Model

A classical decision problem that can be approached using the *Multi-Attribute Decision Model* (MADM) is expressed by means of a $m \times n$ *decision matrix*:

$$\begin{array}{ccccc} & C_1 & C_2 & \cdots & C_n \\ X_1 & x_{11} & x_{12} & \cdots & x_{1n} \\ X_2 & x_{21} & x_{22} & \cdots & x_{2n} \\ \vdots & \vdots & & & \vdots \\ X_m & x_{m1} & x_{m2} & \cdots & x_{mn} \end{array}$$

The lines of the *decision matrix* stand for m decision alternatives $X_1, ..., X_m$ that are considered within the problem and the columns have the significance of n *attributes* or *criteria* according to which the desirability of an alternative is to be judged. An element x_{ij} expresses in numeric form the consequence of the decision alternative X_i with respect to the criterion C_j.

The aim of MADM is to determine an alternative X^* with the highest possible degree of overall desirability. To find the most desirable alternative the decision maker uses an *aggregation approach* which generally has two steps (Zimmermann, 1996):

1. For each decision alternative X_i, the aggregation of all judgements with respect to all goals.
2. The rank ordering of the decision alternatives according to the aggregated judgements.

Within a MADM, the decision maker could express the criteria C_j as linguistic terms, which leads to the utilisation of fuzzy sets to represent them. Thus, as Zimmermann (1996) suggested, is achieved a form of *Fuzzy MADM* and the elements x_{ij} of the decision matrix become values of the corresponding membership functions.

1.3 The aggregation and ranking of the decision alternatives

A simple and frequently used method to perform the aggregation for each decision alternative X_i , is the calculation of an overall rating R_i:

$$R_i = \sum_{j=1}^{n} w_j x_{ij}, \quad i = 1,...,m \qquad (1)$$

where w_j are subjective weights expressing the importance of the criteria to the decision maker.

For the evaluation of the subjective weights w_j, Saaty (1980) proposed a quite simple and intuitive method: *AHP* (*Analytic Hierarchy Process*) which is based on a pairwise comparison of the n criteria. The decision maker builds a $n \times n$ matrix whose a_{kl} elements are assigned with respect to the following two rules:

1. $a_{kl} = \dfrac{1}{a_{lk}}, \quad k = 1,...,n, \; l = 1,...,n;$
2. If criterion k is more important then criterion l, then assign to criterion k a value form 1 to 9.

Table 1 Saaty's scale of relative importance

Intensity of relative importance	Definition
1	equal importance
3	weak importance
5	strong importance
7	demonstrated importance
9	absolute importance
2, 4, 6, 8	intermediate values

A guide (Saaty, 1980) for assigning values in the matrix is shown in Table 1.

The weights w_j are then determined as the components of the eigenvector corresponding to the largest eigenvalue of the above $n \times n$ matrix. An alternative solution that gives a good approximation of the eigenvector components, but is easier to use, consists in calculating the normalised geomean of the lines (Fuller, 1996).

After the computation of the ratings R_i, $i = 1,..., m$, the only operation that remains to be done is to sort the decision alternatives following the descending order of R_i.

In the next section are described the steps of a methodology proposed to be followed for applying the Fuzzy MADM model in the selection phase of a plant breeding program. Concomitantly is presented an application of the proposed methodology.

2. APPLYING THE FUZZY MADM IN THE SELECTION PHASE OF A PLANT BREEDING PROGRAM

Every plant breeding program involves a phase of artificial selection in which the specialist separates the elite from a number of individuals. A number of quantitative characters are observed and measured to make the selection. The result of these observations is a list of numeric values: one row for each plant and one column for each observed character. The list must be examined in order to decide which plants to include in the elite. Usually, a great number of individuals are examined and quite many characters are measured. This is why a decision support tool would be useful and the nature of the problem leads to the utilisation of Fuzzy MADM. The fuzzy techniques are appropriate because the specialist often deals with linguistic values.

The proposed methodology consists in seven steps which are described further on. The application of each step is illustrated through an example that consists in selecting the elite from a set of 150 plants of pod bean, variety Cape. The study was initiated (Leonte *et al.*, 1997) in view of cultivating this kind of pod bean in Romania.

Ten characters were studied for each plant: height, number of branches, number of pods, average length of pods, average diameter of pods, number of beans, average number of beans per pod, weight of beans, average weight of beans per pod, weight of 1000 beans.

Table 2 contains the results obtained for only 10 out of 150 plants and only 5 out of 10 studied characters. It suggests the difficulty of dealing with the entire list of 1500 numerical data.

Table 2 Results of measuring some characters of 10 plants of the pod bean variety Cape

ID	Height (cm)	Branches	Pods per plant	Avg. pods length (cm)	Avg. pods diameter (cm)
1	29	5	10	9.80	0.78
2	43	7	8	8.12	0.68
3	35	7	13	9.84	0.76
4	39	6	12	9.25	0.75
5	41	10	25	11.04	0.73
6	46	8	18	10.33	0.75
7	32	8	17	8.64	0.64
8	31	6	7	9.85	0.85
9	38	7	16	8.18	0.56
10	32	7	10	10.00	0.88

2.1 Defining the linguistic variables

One method that can be used to define linguistic variables is using the information obtained through statistical study. Such studies are always done within plant breeding programs.

The following example describes the fuzzy variable *length_of_pods* that was defined by examining the results of the statistical study. The fuzzy variable regards the (average) length of pods per plant and comprises three fuzzy terms: *low*, *medium*, *high*.

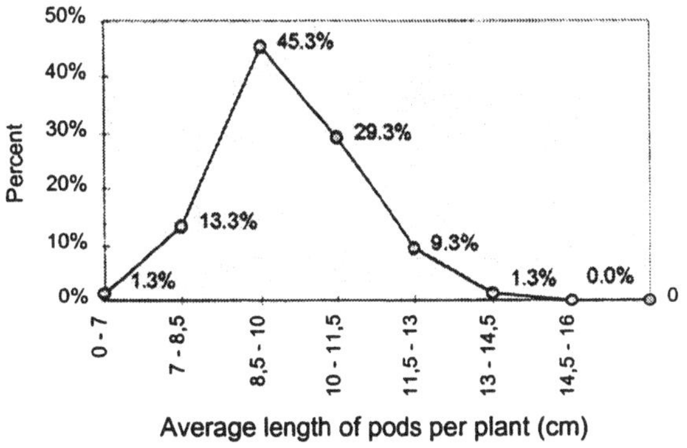

Fig. 1. The variability of the average length of pods per plant

In Figure 1 is shown the variability chart of the character *average length of pods per plant* (cm). Figure 2 shows the corresponding fuzzy variable *length_of_pods*.

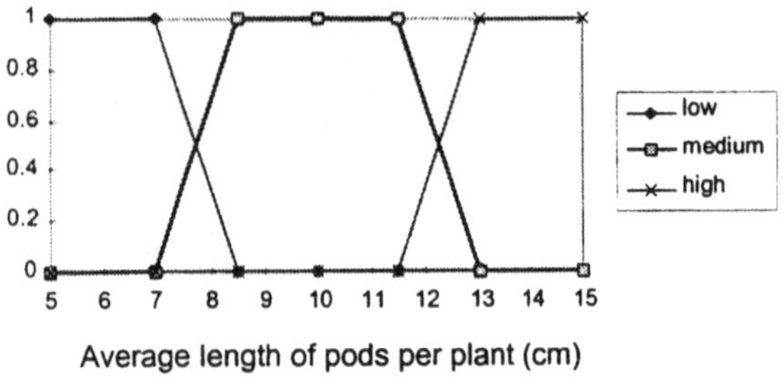

Fig. 2. Representation of the fuzzy variable *length_of_pods*

For each observed character such linguistic variables must be defined even though the plant breeding expert will not use them all. Within this step, an efficient co-operation between the IT specialist and the plant breeding expert is essential.

In the case study the ten fuzzy variables were defined using trapezoidal fuzzy sets. The maximum number of fuzzy terms within a fuzzy variable was five.

2.2 Establishing the selection criteria

The selection criteria would be established using, according to the situation, fuzzy terms or crisp criteria. However, as the case study shows, there could be criteria that seem to have a crisp definition, but must be also fuzzified.

The selection criteria used for selecting the plants in the case study are enumerated further on.

Height: normal. This linguistic value was chosen because the statistical study didn't reveal some special correlation between height and the other characters.

Number of branches: great. A great number of branches determines a great number of pods.

Number of pods: great. Being a pod bean variety, this is the most important criterion.

Average length of pods: medium. Further processing requirements and marketing reasons imposed this linguistic value.

Average diameter of pods: medium. Same reasons as before.

Number of beans: medium. Even though it's a pod bean variety, beans are important for the nutritional and taste qualities.

Average number of beans per pod: medium. Same reasons as before.

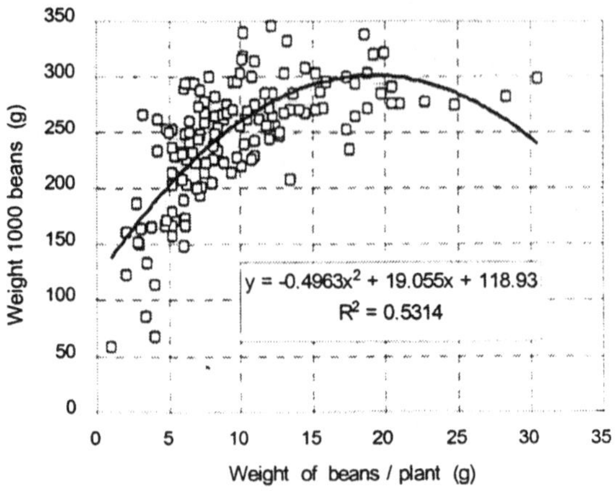

Fig. 3. Correlation between the weight of beans per plant and the weight of 1000 beans

Weight of beans: approximately 15 - 20 g. The choice of this criterion as an interval was determined by the results obtained in studying the correlation between this character and the others. Figure 3 shows such a correlation.

However, a fuzzification of this criterion was also made to allow degrees of variable confidence on the both sides of the interval. The corresponding trapezoidal fuzzy set is shown in Figure 4.

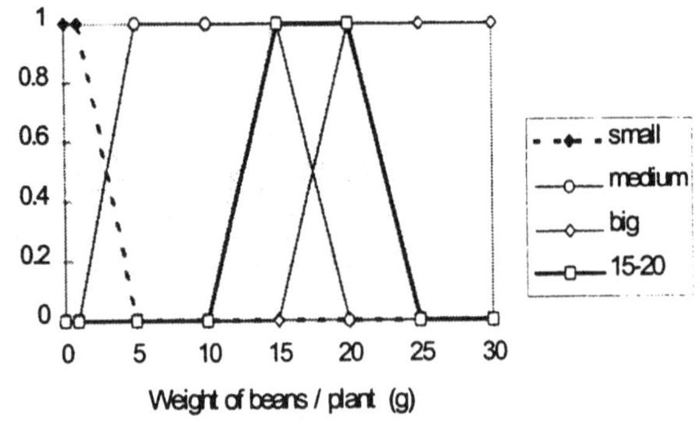

Fig. 4. Fuzzification of the interval 15-20

Average weight of beans per pod: approximately between 1 and 1.5 g. Same reasons as before determined the choice of a fuzzified interval.

Weight of 1000 beans: big. This parameter ensures a high quality of beans; this is why big values are desirable.

2.3 Applying the selection criteria

The results of applying the selection criteria expressed as fuzzy sets, are values of the corresponding membership functions, that is the measure of satisfaction of each criterion by each plant.

Table 3 Results of applying the selection criteria on some characters of 10 plants of the pod bean variety Cape

ID	Height (cm)	Branches	Pods per plant	Avg. pods length (cm)	Avg. pods diameter (cm)
1	0.33	0.	0.	1.	1.
2	0.	0.	0.	0.75	1.
3	0.67	0.	0.	1.	1.
4	0.	0.	0.	1.	1.
5	0.	1.	0.	1.	1.
6	0.	0.	1.	1.	1.
7	1.	0.	1.	1.	1.
8	1.	0.	0.	1.	0.50
9	0.	0.	1.	0.79	0.60
10	1.	0.	0.	1.	0.20

For the studied plants some results are shown in Table 3. The same plants and same characters as in Table 2 were chosen.

2.4 Calculating the weights of the selection criteria using Saaty's method

One of the most time consuming steps of this methodology is the pairwise comparison of the selection criteria in view of determining the elements of Saaty's matrix, defined in section 1.3. In an early stage it would be easier to perform this step by interview. Thus, the interdisciplinary co-operation between the IT specialist and the plant breeding expert is again of greatest importance. In Table 4 is shown a block of 5 × 5 elements of the Saaty's matrix, calculated for the selection criteria in the case study.

Table 4 Elements of the Saaty's matrix (5 × 5 block)

	Height	Branches	Pods/ plant	Avg. pods length	Avg. pods diam
Height	1	1/3	1/7	1/5	1/5
Branches	3	1	1/5	1/3	1/3
Pods / plant	7	5	1	4	1
Avg. pods length	5	3	1/4	1	1
Avg. pods diameter	5	3	1	1	1

Following the procedure described in 1.3 the final part of this step is the computation of relative weights that give the importance of each criterion.

The computed values of the relative weights in the case study are listed in Table 5. As one can see, the most important character is the number of pods per plant. On the next two places are other important characters: the diameter of pods and the length of pods. Hence, the calculus confirms some obvious intuitive ideas.

Table 5 The weights expressing the relative importance of the studied characters

Character	Weight
Height	0.105022
Number of branches	0.399718
Number of pods	1
Average length of pods	0.584892
Average diameter of pods	0.736336
Number of beans	0.168555
Avg. number of beans per pod	0.354378
Weight of beans	0.180509
Average weight of beans per pod	0.218489
Weight of 1000 beans	0.171152

2.5 Computing the overall ratings for each decision alternative

The overall ratings are computed according to relation (1) defined in section 1.3. Thus, to each studied plant will be assigned a final score. The maximum possible score would be equal to the sum of the weights.

For the studied pod bean variety this calculation involves the membership degrees shown partially in Table 3 and the weights shown in Table 5. Relation (1) becomes

$$R_i = \sum_{j=1}^{10} w_j x_{ij}\ , \quad i = 1,\ldots,150 \qquad (2)$$

The maximum score which one plant can obtain is, in the case study, equal to 3.919442.

2.6 Ranking the studied individuals

The final hierarchy is determined by sorting the overall ratings in descending order.

The top ten final results for the case study are listed in Table 6. On the first place is the individual having ID number 110 which attained a final score of 3.57, that is 91.12% of the maximum possible score.

Table 6 The final hierarchy determined by sorting the overall ratings in descending order (partial)

Place	ID	Score	Percent
1	110	3.57	91.12%
2	39	3.37	85.86%
3	28	3.01	76.81%
4	117	2.95	75.16%
5	64	2.64	67.43%
6	31	2.63	67.04%
7	96	2.61	66.59%
8	21	2.57	65.55%
9	149	2.52	64.34%
10	51	2.51	64.16%

2.7 Selecting the elite

The plant breeding expert can choose as many individuals as he considers to be appropriate, from the top of the final classification. A graphical representation of the results can help in making a good decision. Such a diagram is shown in Figure 5 for the Cape pod beans.

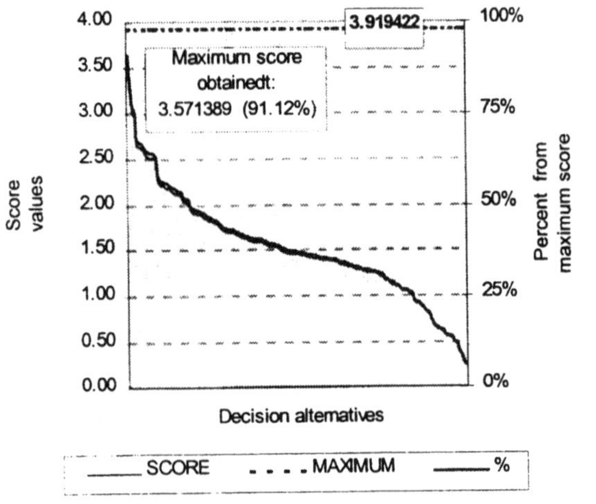

Fig. 5. The diagram showing the final ranking of the decision alternatives

CONCLUSION

The selection phase of a plant breeding program can be viewed as a Fuzzy Multi-Attribute Decision problem. A methodology to apply this model was proposed.

The construction of an user friendly decision support software could help the plant breeding expert to improve the quality of his decision act. Implementation of fuzzy techniques would permit him to deal with linguistic qualifiers.

The results of the statistical study performed on the sets of plants involved can be used to define the fuzzy variables. Such statistical studies are present in all plant breeding programs.

Some selection criteria could be expressed by intervals instead of linguistic values. However these intervals would be transformed in fuzzy sets to ensure a better utilisation.

The training of the plant breeding specialists in using the fuzzy sets concepts was not difficult. The ranking of the importance of the selection criteria using the Saaty matrix was made correctly.

REFERENCES

Calin, M., C. Leonte and S.C. Buraga (1998). Applying Fuzzy Techniques for Representing Knowledge in Horticultural Research (published in Romanian). In: *Advances in the Romanian Horticultural Research* (A. Gherghi ed.). Medro, Bucharest, Romania.

Fuller, R. (1996). Fuzzy Decision Making, Lecture notes. *www.tucs.abo.fi/courses/95-96/material/fuzzydec.html.*

Leonte, C., G. Târdea and M. Calin (1997). On the correlation between different quantitative characters of Cape pod bean variety (published in Romanian). *Lucrari stiintifice, Horticultura*, USAMV Iasi, Romania, **Vol. 40**, 136-142.

Reusch, B., (1996). Mathematics of Fuzzy Logic. In: *Real World Applications of Intelligent Technologies* (H-J. Zimmermann, M. Negoita, D. Dascalu eds.). Publishing House of the Romanian Academy, Bucharest, Romania.

Saaty, T.L. (1980). *The Analytic Hierarchy Process: Planning, Priority Setting, Resource Allocation.* McGraw-Hill, New York.

Zimmermann H-J., 1996 - Fuzzy Decision Support Systems, In: *Real World Applications of Intelligent Technologies* (H-J. Zimmermann, M. Negoita, D. Dascalu eds.). Publishing House of the Romanian Academy, Bucharest, Romania.

GREENHOUSE CLIMATE CONTROL FOR INTEGRATED PEST MANAGEMENT

Hans-Juergen Tantau[1)] and **Doris Lange**[2)]

[1)]*Institute of Horticultural and Agricultural Engineering,
University of Hannover, Herrenhaeuser Str. 2, D-30419 Hannover,
e-mail:* tantau@itg.uni-hannover.de
[2)] *Landwirtschaftskammer Westfalen-Lippe - Referat Gartenbau,
Nevinghoff 40, 48147 Münster*

Abstract: For greenhouse climate control modern control strategies are used in order to influence plant development as well as the reduction of disease infestation. Due to these considerations, it is necessary to minimize chemical applications, within the concept of integrated plant protection. In addition to existing dehumidification strategies, in particular the use of climate control for integrated pest control requires registration and simulation of climate conditions in the plant canopy . To avoid intensive and expensive technical measurements, it is necessary to describe the energy and mass transport processes within the canopy, the exchange processes between air and plant elements and other surfaces. Based on a plant canopy model the basics for of a computer supported anti-botrytis climate-control management have been developed.

Keywords: computer control, modelling, computer simulation, integrated plant control

1. INTRODUCTION

During the past decade Integrated Pest Management (IPM) has become a well-known proper method to control pests more efficiently. The major goal of IPM is to maintain pest populations below economic thresholds while utilising suitable techniques to protect both the environment and non-target species. This goal can be achieved by modelling the initial population, the host and the environment. Using the integrated approach to pest management requires to study the interactions between the physiology and biochemistry of the plant and the pest, as well as the interactions between both organisms and the microenvironment. As the plants can profoundly affect the canopy microclimate, and vice versa, it becomes essential to use an integrated approach that considers simultaneously most of the important plant and environment variables. The microenvironment variables may be important to the success or failure of a particular pest.

The achievements in the climate study did provide good information on the conventional greenhouse environmental control, which enables plants to be grown in economically way. Most of the studies treated the canopy as a uniform plane and greenhouse air as an uniform block. Multi-layer simulation models have been proposed in literature to figure out radiation penetration and vertical profiles of temperature and humidity intensity (Goudriaan, 1977; Myeni and Impens, 1985; von Elsner, 1982; Chen, 1984; Yang, 1988). However, they rely on time consuming calculations and on a huge amount of input parameters. For the purpose of an IPM system controlling water vapour and temperature in greenhouses must end up in <u>one</u> 'mean' value of temperature and water vapour within the canopy.

In this respect, the energy balance method is applied to a canopy, using basic physical and physiological principles to describe the exchange and transport processes and to predict canopy temperature and water vapour concentration under different systems. The relationship between greenhouse climate, water vapour production and canopy microclimate will be analysed by a set of definitions and a step-by-step approach. The experimental determination of the parameters was performed by studying the influence of different heating and production systems and plant densities on canopy microclimate. In addition disease incident of grey mold, caused by *Botrytis cinerea* Pers. was measured during production period. Secondly, these measurements were compared with a simple pathogen model for *Botrytis cinerea* Pers., using canopy temperature and saturation deficit as input variables for simulation of pathogen development rate.

2. THEORETICAL BACKGROUND

2.1 *Physical model of the plant canopy*

A physical model has been developed to describe the heat and mass transfer between the canopy and the air in the greenhouse. The canopy is regarded as an homogenous layer, characterised by one value of temperature and vapour pressure. This layer is bounded by an ideal surface within the greenhouse, which is also homogenous (fig. 1). The energy exchange between the crop and its environment is governed by conduction, convection of sensible and latent heat and thermal radiation. In addition, energy is partly converted to products of photosynthesis. The general energy balance is represented with:

$$R_n = H + LE + M + J \qquad (1)$$

Because of its small magnitude, the effects of energy stored in dry matter production and the thermal storage will be neglected. Following Monteith (1990), the concept of resistance was adopted to

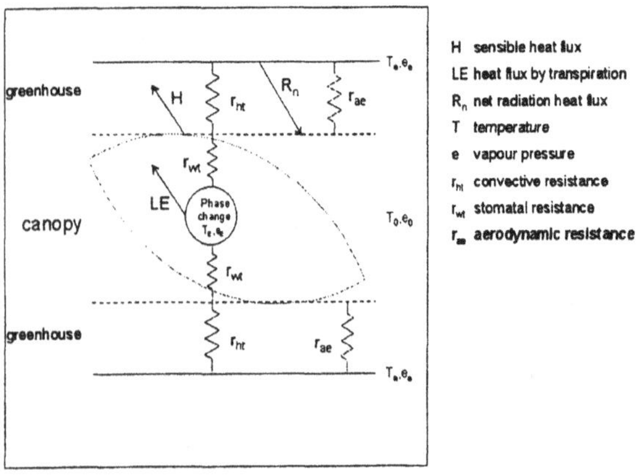

Fig. 1. Physical model for heat and mass transfer at the canopy.

formulate the heat and mass transfer at the leaf surface.

Energy and mass balances

The energy fluxes at the upper and lower surfaces of a canopy can be defined as the balance of latent and sensible heat released by the canopy. The sensible heat flux of a single leaf with an exchange area of 2 LAI can be described with:

$$H = \frac{2 \cdot LAI \cdot \rho c_p}{r_{ht}} (T_0 - T_a) \qquad (2)$$

In an analogous way the canopy transpiration rate is defined by the Penman-Monteith equation with:

$$LE = \frac{\frac{\delta}{\gamma}(R_n - J) + \frac{2 \cdot LAI \cdot \rho c_p}{\gamma r_s} \rho c_p (e_a^* - e_a)}{1 + \frac{\delta}{\gamma} + \frac{r_i}{r}} \qquad (3)$$

and temperature difference:

$$T_o - T_a = \frac{\frac{r_{wt} + r_{ht}}{\rho c_p} \cdot R_n - \frac{1}{\gamma} \cdot (e_a^* - e_a)}{1 + \frac{\delta}{\gamma} + \frac{r_{wt}}{r_{ht}}} \qquad (4)$$

Net radiation regime of plant canopy

The net radiation regime is defined as the total energy intercepted and absorbed by plant elements. The plant stand is modelled as a horizontal homogenous and optically isotropic turbid medium. The exchange of short- as well long- wave radiation is developed as a function of optical and geometrical properties of the canopy and underlying surface and leaf area index (LAI), where the radiation absorbed by the canopy (R_n) is calculated with:

$$R_n = (1 + \tau_s \rho_g) \cdot (1 - \tau_s - (1 - \tau_l)\rho_\infty) I_s + (1 + \tau_l) \cdot (I_{l,g} + I_{l,c} - 2\sigma T^4) \qquad (5)$$

The fraction of radiation reflected by the underlying surface and intercepted by the lower part of the canopy is represented by $\tau_s \rho_g$. The definition equation for long- and short-wave transmittance are given by:

$$\tau_l(LAI) = e^{-k_l LAI} \qquad (6)$$

$$\tau_s(LAI) = e^{-k_s LAI} \qquad (7)$$

$$\rho_s(LAI) = (1 - \tau_l(LAI))\rho_\infty + (r_s^2(LAI))\rho_g \qquad (8)$$

Long- and short-wave transmittance τ_l and τ_s and the parameters ρ_∞ and ρ_g were empirically determined, since they embody the influence of the crop peculiarities on the transfer of radiation. The defining equations for the short- and long- wave transmittance were substituted in the general equation for net radiation with the experimentally determined coefficients and geometrical coefficients, deducted from the geometry of the present greenhouse to the thermal radiation emitted by the heating pipes and the underlying surface.

Convective heat transfer

Transfer of sensible heat across an air layer takes place through an exchange area equal to the leaf area (of both sides), across a total plant resistance to sensible heat transfer (r_{ht}), assuming that all leaves in the canopy act in parallel. The transfer resistance can be expressed by the non-dimensional Nu number in the following form:

$$r_{ht} = \frac{c_p \cdot \rho \cdot \ell}{\lambda_a \cdot Nu} \quad (9)$$

Assuming that energy and mass exchange of the plant elements are governed by processes described as mixed convection, the empirical relation for heat transfer with combined effect of laminar, forced and free convection, was adopted to describe convective heat transfer between the canopy and the air as:

$$Nu = 0.68 \cdot \mathrm{Re}^{0.5}\left(\left(\frac{Gr}{\mathrm{Re}^2}(0.958 + \mathrm{Pr})\right)^{0.5} + \mathrm{Pr}\right)^{0.5} \quad (10)$$

Plant responses

Two very important physiological processes of plants, that respond on environmental factors, are transpiration and photosynthesis. Applying the analogue resistance sub-model introduced by Monteith (1973), these processes are formulated as schematically shown in figure 1. The transfer resistance (r_{wt}) of water vapour is impeded by the stomatal (r_s) and cuticular (r_c) resistances, represented with:

$$r_{wt} = \frac{r_s \cdot r_c}{r_s + r_c} \quad (11)$$

The stomatal resistance to water vapour transfer, derived on the basis of physical principles is related to radiation (PAR) and CO_2 concentration. For leaves in their natural environment r_s strongly depends on the photosynthesis. In absence of light stomata are usually closed, so that transpiration is effectively zero. This is described with:

$$r_s = \frac{\frac{C_i}{C_a} \cdot CO_2}{\frac{D_{wt}}{D_{CO2}} \cdot P \cdot V_m} \quad (12)$$

and:

$$P = \beta(T)\frac{P_{\max}}{k_s} \cdot \frac{(1-\tau_s)P_{\max} + q_e I_{PAR} \cdot k_s}{(1-\tau_s)P_{\max} + q_e I_{PAR} \cdot k_s \cdot A_s} \quad (13)$$

$$A_s = (1 + \tau_s \rho_g) \cdot (1 - \tau_s - (1-\tau_l)\rho_\infty) \quad (14)$$

Cuticular resistance (r_c) is generally acknowledged to be large and certainly comparable with the resistance of closed stomata.

Water vapour concentration of the canopy

After quantifying the canopy temperature and the transpiration rate, the absolute humidity of the leaf surface, deducted from eq. (3) is given by:

$$X_{eff} = X_a^* + \frac{\delta}{\gamma}\frac{r_{ht}}{2LAI}\frac{R_n}{L} \quad (15)$$

The concentration gradient $X_{eff} - X_a$ is the driving force for transpiration. The water vapour concentration of the ambient is balanced by the vapour fluxes of transpiration, condensation and ventilation, with a vapour concentration at equilibrium given by X_a. Representing the canopy as a porous medium implies an homogenous ambient of the greenhouse and within the canopy. To overcome this limitation, the water vapour concentration between the ambient air of the greenhouse and the canopy is described in a linear form between the limitations:

$$X_0 = X_a^* + \frac{\delta}{\gamma}\frac{r_{ht}}{2LAI}\frac{R_n}{L} \quad \text{for a dense stand} \quad (16)$$

$$X_0 = C\frac{e_a}{T_o} \quad \text{for } LAI \le 0.5 \quad (17)$$

The structure of the canopy model is shown in fig. 2.

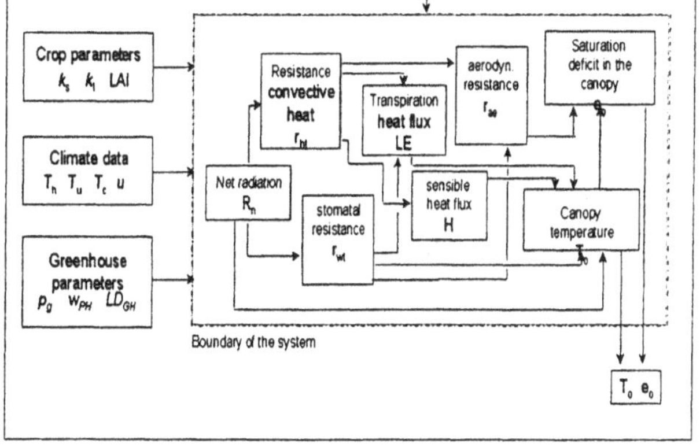

Fig. 2. Structure of the canopy model.

2.2 *Prediction of development rate of Botrytis cinerea Pers.*

Disease caused by the ubiquitous fungus *Botrytis cinerea* was calculated as development rate with a combined function of temperature and saturation deficit. The non linear relationship between temperature, saturation deficit and development rate are based upon general descriptions by Analytis (1977) and Friedrichs (1994). The calculations were started from the same phenological state of the plants, the peel off of the epidermis by thickening, using hourly values of climatic variables.

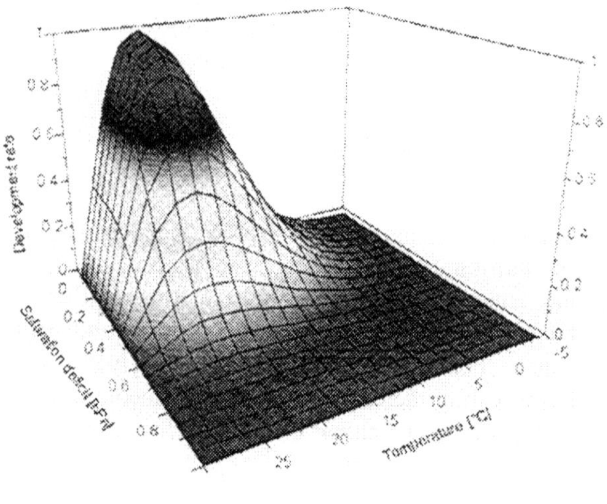

Fig. 3. Development rate of *Botrytis cinerea* depending on temperature and saturation deficit.

2.3 Measurements

The Measurements were carried out during early spring season at the Research station for Horticulture, Hanover. The glasshouses were covered with single glass. Thermal screen and different heating systems are available in combination with commercial white plastic benches and aluminium gutters.

The crop (*Fuchsia x hybrida* 'Beacon') was grown in 11 cm plastic pots, which were placed to give a LAI from 1 to 8 during cropping period.

Incoming solar radiation was measured by solarimeter (Kipp and Zonen, Delft, Netherlands), placed 1 m above the canopy. Wet and dry bulb temperature were measured in 0.7 m above the canopy by aspirated Psychrometer. In order to avoid air movements, the humidity in the canopy was measured with capacitive sensors (Testo, Lenzkirch, Germany). Temperatures of heating pipes, of benches and aluminium gutters were measured with thermocouples (NiCr-Ni, type K) glued to the surface. The temperature of the leaves was measured by thin thermocouples (0.1 mm, NiCr-Ni) touching the lower surface of the leaves facing north. Two net radiometers (Schenk, Wien, Austria) were installed above the canopy. All data's were measured at an interval of 15 s, using a data logging system (ITG, Hanover, Germany). Every 10 minutes average values were stored on disk for further evaluations.

The growth of lesions was scored weekly on 30 plants, using the following scoring system in table 1.

Table 1. Scoring system

Score class	
1	Lesions on 1-2 shoots
3	Lesions on 3-5 shoots
5	Lesions on >6 shoots
7	Total break down

3. RESULTS AND DISCUSSION

The experimental determination of the parameters eq. (5) resulted to be 1.12 for the mean extinction coefficient for short-wave radiation (k_s). Assuming reflectance and transmittance values for leave tissue with $\rho_t = 0.30$ and $\tau_t = 0.2$ (Ross, 1975) yields $k_l = 0.75$. Based on the extinction coefficient for horizontal leaf angle distribution, the albedo of the canopy is assumed to be 0.093 (Goudriaan, 1977). The net radiation was estimated through eq. (5) by fitting the parameters in the equation. Comparing with measured net radiation flux there is a sufficient agreement between the two estimates ($r^2 = 0.94$), proving that this method is qualified for estimation of net radiation from a canopy.

In figure 4 (upper part) the daily course of the canopy temperature is displayed (calculated vs. observed) for two consecutive days. In general plant temperature, calculated by the model reproduces quite satisfactory the measured temperatures. Although in the shown example plant temperature is closed to air temperature it was observed that during heating

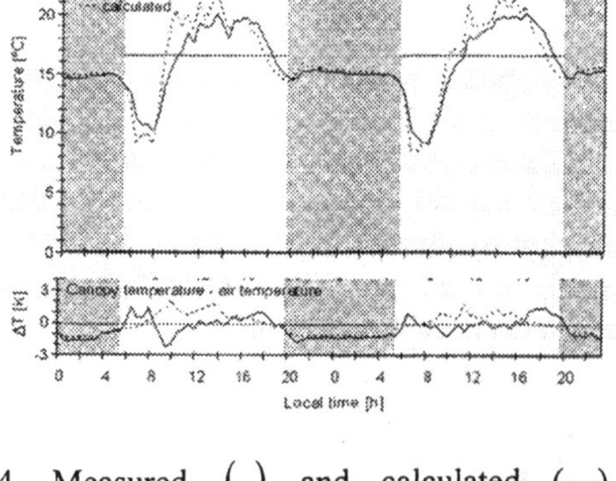

Fig. 4. Measured (–) and calculated (---) plant temperature to air temperature excess for two days period.

period with overhead heating pipes plant temperature was below air temperature. In contrast a heating system installed under the gutter system leads to an increased plant temperature. Comparing the temperature excess of measured and calculated foliage temperature to air temperature (fig. 4, lower part) clarifies that temperature of the plant surface is not equal to air temperature. Dealing with the humidity of the canopy it is not enough to state that water vapour concentration or saturation deficit of the ambient temperature is the natural output of the vapour balance of the greenhouse air. In figure 5 the diurnal course of measured (–) and calculated (---) saturation deficit based on water vapour concentration of the ambient and on estimation by eq. (16) and (17) is shown for the period of two days. With increasing LAI vapour exchange of the canopy and the ambient air is impeded reaching nearly saturation for close stand.

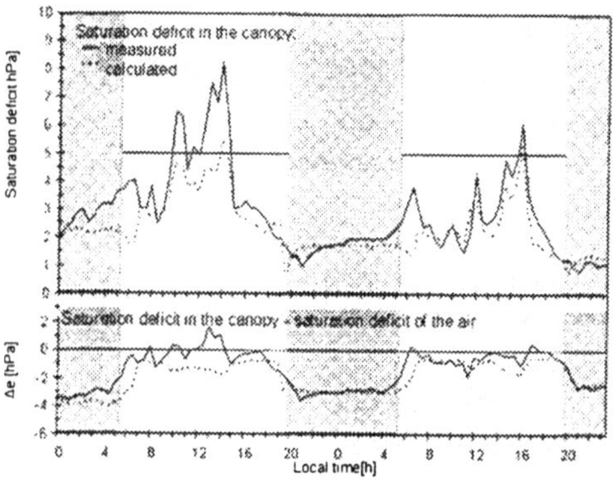

Fig. 5. Measured (–) and calculated (---) saturation deficit at plant temperature based on measured ambient vapour concentration and calculated through eq. (16) and (17).

Figure 6 outlines the results of the epidemiological investigations at flowering state of the plants. The number of infections with *Botrytis cinerea* is shown as a function of control strategies, irrigation system and plant density. The results make clear that the plant density is a very important parameter for integrated disease control. Second important is the irrigation system. If the parameters for plant density and for the irrigation system are optimal then greenhouse climate control can be used efficiently to reduce the risk of infection. The fact that the mean temperature of the plant differs from that of the air and thus produces a quite different saturation deficit has important implications for the disease incidence. In order to develop an Integrated Pest Management (IPM) system for greenhouse crops, the fitting of a canopy model into a general greenhouse model provide detailed information on climatic variables within a plant canopy to derive specific requirements for control operations and/or fungicide applications.

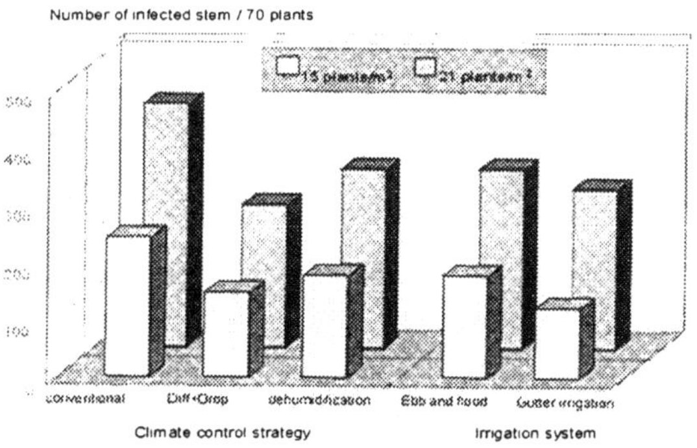

Fig. 6. Infections with *Botrytis cinerea* as a function of control strategies, irrigation system and plant density

In figure 7 the set up for the integrated pest management (IPM) is shown. For the application a few parameters for the greenhouse, the crop and for the disease are necessary

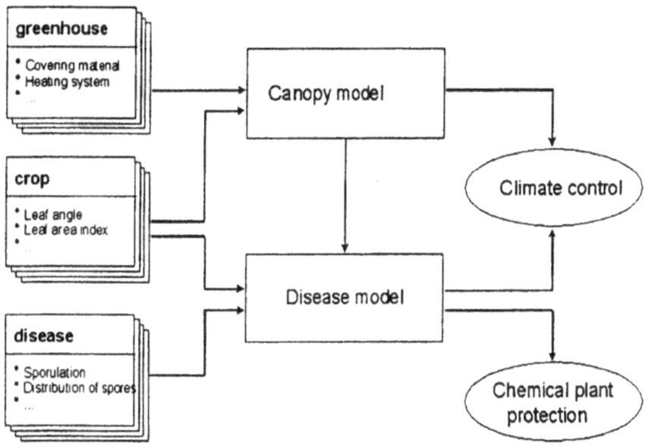

Fig. 7. Set up for the integrated pest management

Using this set up integrated pest management is possible by greenhouse climate control. As mentioned earlier the effect depends on other parameters - mainly plant density and irrigation system . If the risk of infection is too high the recommendation for chemical disease control must be given by the system.

It is planned to test the implementation into a climate computer on four ornamental farms.

REFERENCES

Analytis, S. (1977). Über die Relation zwischen biologischer Entwicklung und Temperatur bei phytopathogenen Pilzen. *Phytopath.Z.*,90, 64-76.

Chen,J. (1984). Mathematical analysis and simulation of crop micrometeorology. *Ph.D. Dissertation.* Agricultural University, Wageningen.

Elsner, B.v. (1982). Das Kleinklima und der Wärmeverbrauch von geschlossenen Gewächshäusern. *Ph.D. Dissertation* TU Hannover.

Friedrich, S. (1994). Prognose der Infektionswahrscheinlichkeit durch echten Mehltau an Winterweizen (*Erysiphe graminis* DC f. sp. tritici) anhand meteorologischer Eingangsparameter. *Ph.D. Dissertation.* TU Braunschweig.

Goudriaan,J. (1977). Crop micrometeorology: a simulation study. *Simulation Monographs.* Pudoc. Wageningen.

Lange,D. (1999). Bestandsklimamodell zur Anwendung im integrierten Pflanzenschutz am Beispiel von *Botrytis cinerea* Pers. *Ph.D. Dissertation* University of Hannover.

Monteith,J.L. (1973). Principles of Environmental Physics. *Edward Arnold.* London.

Myneni,R.B. and I. Impens (1985). A procedural approach for studying the radiation regime of infinite and truncated foliage sparces. Part II. Experimental results and discussion. *Agric. For. Meteorol.,* 34, 3-16.

Ross,J. (1975). Radiative transfer in plant communities. In: Vegetation and Atmosphere (J.L. Monteith, Ed.), pp 13-55. *Academic Press.* London.

Yang,X. (1988). Greenhouse microclimate. Transport processes, plant responses and dynamic modelling. *Ph.D. Dissertation.* The Ohio State University. Columbus, Ohio.

APPENDIX

The following symbols are used:

A	absorption coefficient	-
c_p	specific heat content	J kg^{-1} K^{-1}
CO_2	ambient CO_2 concentration	vpm
D	diffusion coefficient	m s^{-1}
E	evaporation flux density	kg m^{-2} s^{-1}
e	vapour pressure	Pa
Gr	Grashof number	-
H	flux density of sensible heat	W m^{-2}
*	at saturation	
I	irradiance	W m^{-2}
I_{PAR}	light energy flux	µmol m^{-2} s^{-1}
J	rate of thermal storage	W m^{-2}
k	extinction coefficient	-
L	latent heat vaporisation	J kg^{-1}
LAI	leaf area index	-
LE	latent heat flux	W m^{-2}
M	photosynthetic rate	W m^{-2}
Nu	Nusselt number	-
P	photosynthesis	µmol m^{-2} s^{-1}
Pr	Prandtl number	-
q	leaf quantum efficiency	-
R	radiation flux density	W m^{-2}
Re	Reynolds number	-
r	transfer resistance	s m^{-1}
T	temperature	K
V_m	molecular volume of a gas	m^3 mol^{-1}
X	Concentration	kg m^{-3}
β	temperature reduction factor	-
γ	psychometric constant	Pa K^{-1}
ρ	surface reflectance	-
ρ	density	kg m^{-3}
σ	Stephan Boltzmann constant	Wm^{-2}K^{-4}
τ	transmittance	-

subscripts

a	ambient air
b	boundary layer
c	of the greenhouse cover
c	cuticular
CO_2	of carbon dioxide
E	at phase-interface
g	underlying surface
h	pipe heating system
ht	heat
l	long-wave
max	maximal
n	net
0	at external surface
q	cuticular
s	short-wave
s	stomatal
t	leaf tissue
wt	water vapour
∞	of a dense stand

superscripts

A NON-LINEAR FEEDBACK TECHNIQUE FOR GREENHOUSE ENVIRONMENTAL CONTROL

G.D.Pasgianos[1], K.G.Arvanitis[2], N.A.Sigrimis[2]

[1] *National Technical University of Athens, Dept. of Electrical and Computer Engineering, Zographou 15773, Athens, GREECE*
[2] *Agricultural University of Athens, Dept. of Agricultural Engineering, Iera Odos 75, 11855, Athens, GREECE, email: karvan@aua.gr, n.sigrimis@computer.org*

Abstract: A new approach to system linearization and decoupling is presented for the operation of heating/cooling and moisturizing of greenhouses. High level programming, such as the virtual variables which provide an easy way to building models, is a feature of most research but also field control systems. The method is applicable to any air-conditioning system and is expected to be gain wide acceptance in modern SCADA systems with extended computational capabilities. *Copyright© 2001 IFAC*

Keywords: Greenhouse climate, environmental control, psychrometrics, feedback linearization, nonlinear systems

1. INTRODUCTION

Several studies and research applications involving environmental control of greenhouses have been performed by many researchers (Jones *et a.l*, 1984; Gates and Overhults, 1991; Stanghellini and van Meurs, 1992; Zhang and Barber, 1993; Stanghellini and de Jong, 1995; Chao *et al*, 1995; Chao and Gates, 1996; Arvanitis *et al.*, 2000; Chao *et al.*, 2000; Zolnier *et al*, 2000). Most of the studies on analysis and control of the environment inside greenhouse have been based on the concept of energy and mass balance and physical modeling. These concepts are very effective in order to clarify the concepts of environmental control, to refine environmental control strategies, and to gradually lead to economic optimization, the ultimate objective of environmental control.

Many dynamic models for greenhouse environment exist in the extant literature, and they are of nonlinear nature. The central state variable is typically air temperature, with relative humidity (or absolute humidity) and carbon dioxide concentration also considered. Disturbances to a greenhouse or other plant thermal environment occur primarily from solar radiation, outside temperature (conduction heat transfer and ventilation heat transfer), and interactions with occupants (plants), the controlled heating and ventilating equipment, and the floor. However, it is useful to note that for the most part the system is subjected to relatively low frequency disturbances. Indeed, most of these disturbances are considered as "loads" and a quasi-steady state analysis often suffices for design purposes. Perhaps the most common transient disturbance is a step change, either from switching equipment, changing set points, or variable cloud cover.

The fact that temperature and humidity are coupled, and the actuators (i.e. windows) are usually subject to changing characteristics (the gain is largely perturbed by cross product terms with disturbances, such as wind velocity, outside temperature etc) has not been treated analytically to provide a robust control scheme. The practical controllers do meet the control requirements using many expert type of actuator adjustments and ad hoc compensators.

To demonstrate some salient features of greenhouse environmental control, an example of a coupled, nonlinear controller for air temperature and humidity is presented.

2. F/F LINEARIZATION AND DECOUPLING

Consider the analytic nonlinear system

$$\dot{\mathbf{x}} = \mathbf{a}(\mathbf{x},\mathbf{v}) + \mathbf{B}(\mathbf{x},\mathbf{v})\mathbf{u}\ ,\ y_i = h_i(\mathbf{x}) \tag{1}$$

where $\mathbf{x} \in \Re^n$, is the state vector, $u_i, y_i \in \Re$, i=1,...,p, is the ith control input and output, respectively, and $\mathbf{v} \in \Re^d$ is the external disturbance vector. In (1) $\mathbf{a}(\mathbf{x}, \mathbf{v})$, $\mathbf{B}(\mathbf{x}, \mathbf{v})$, and $\mathbf{h}(\mathbf{x})$ are analytic matrix valued functions.

In the case where, system disturbances $\mathbf{v}$ are unknown (or cannot be measured), there is no general theoretical framework, in order to control a system of the form (1). However, in the case where disturbances can be measured, and system (1) can be brought to the form

$$y_i^{(r_i)} = f_i(\mathbf{x},\mathbf{v}) + \mathbf{g}_i^T(\mathbf{x},\mathbf{v})\mathbf{u}\ ,\ i=1,\ldots,p \tag{2}$$

where r_i is the relative degree of the ith system output (Isidor, 1981), then, assuming that matrix $\mathbf{D}(\mathbf{x},\mathbf{v})$ of the form

$$\mathbf{D}(\mathbf{x},\mathbf{v}) = \begin{bmatrix} \mathbf{g}_1^T(\mathbf{x},\mathbf{v}) \\ \vdots \\ \mathbf{g}_p^T(\mathbf{x},\mathbf{v}) \end{bmatrix}$$

is nonsingular, the control law of the form

$$\mathbf{u} = \mathbf{D}^{-1}(\mathbf{x},\mathbf{v})\left\{ -\begin{bmatrix} f_1(\mathbf{x},\mathbf{v}) \\ \vdots \\ f_p(\mathbf{x},\mathbf{v}) \end{bmatrix} + \begin{bmatrix} \hat{u}_1 \\ \vdots \\ \hat{u}_p \end{bmatrix} \right\} \tag{3}$$

where $\hat{u}_i$, i=1,...,p, is a set of external inputs, renders the closed-loop system, I/O linearized, decoupled and disturbance isolated, having the form

$$y_i^{(r_i)} = \hat{u}_i \tag{4}$$

provided that, the system states are measurable.

Note that, in order to bring system (1) in the form (2), it is necessary that, if a disturbance appears in an equation in (1), a control input to be also present in the same equation, allowing elimination of the disturbance by feedforward action. Note that, this feedforward action is inherently present, due to the terms involved in matrices $\mathbf{D}(\mathbf{x},\mathbf{v})$ and $f_i(\mathbf{x},\mathbf{v})$.

Note also that if $\sum_{i=1}^{p} r_i < n$, then, system (1) contains some additional unobser-vable states, called internal dynamics. The *zero-dynamics* of (1) are the internal dynamics of the system when the outputs of the system are kept at *zero* by the input. For the closed system to be stabilizable, the system zero-dynamics must be stable (Isidori, 1981).

Obviously, the closed-loop system (4) can now be controlled by adding an "outer loop" control, in order to satisfy some control specifications. This outer control loop may be based on any conventional linear control strategy, such as pole placement, model matching, H^∞-control, and can be as simple as a PID controller. For example, in pole placement control, application of the outer control law

$$\hat{u}_i = -\sum_{j=0}^{r_i-1} a_{ij} y_i^{(j)} + b_i \tilde{u}_i \tag{5}$$

brings the new closed-loop system to the form

$$y_i^{(r_i)} + \sum_{j=0}^{r_i-1} a_{ij} y_i^{(j)} = b_i \tilde{u}_i$$

Furthermore, in the case of set-point tracking, in order to compensate disturbances, which have not been taken into account in (1) or parametric uncertainties, and in order to attain asymptotic convergence of the error to zero, despite these uncertainty, an additional control loop with integral action (e.g. a PID controller) must be used in most cases. More sophisticated control strategies, such as adaptive controllers, can also be used in some cases.

3. GREENHOUSE VENTILATION MODEL

3.1. Greenhouse dynamic model

The dynamic model energy and mass balance of greenhouse air is shown to be highly nonlinear. A simple greenhouse heating-cooling ventilating model can be obtained by considering the differential equations, which govern sensible and latent heat, as well as water, balances on the interior volume. These differential equations are as follows

$$\frac{dT_{in}(t)}{dt} = \frac{1}{\rho C_p V}\left[q_{heater}(t) + S_i(t) - \lambda q_{fog}(t)\right] - \frac{\dot{V}(t)}{V}\left[T_{in}(t) - T_{out}(t)\right] - \frac{UA}{\rho C_p V}\left[T_{in}(t) - T_{out}(t)\right] \tag{6a}$$

$$\frac{dw_{in}(t)}{dt} = \frac{1}{V} q_{fog}(t) + \frac{1}{V} E(S_i(t), w_{in}(t)) - \frac{\dot{V}(t)}{V}\left[w_{in}(t) - w_{out}(t)\right] \tag{6b}$$

where T_{in} is the interior temperature (°C), T_{out} is the outside temperature (°C), V is the greenhouse volume (m^3), UA is the heat transfer coefficient (WK^{-1}), ρ is the air density (in kg/m^3), C_p is the specific heat of air (J/(kg.K)), q_{heater} is the heat provided by the greenhouse heater (W), S_i is the intercepted solar radiant energy (W), q_{fog} is the water capacity of the fog system (gr/s), λ is the latent heat of vaporization (2257 J/g), $\dot{V}$ is the ventilation rate (m^3/sec), w_{in} and w_{out} are the interior and exterior absolute humidity (absolute water content, g/m^3), respectively, and $E(S_i, w_{in})$ is the evapo-transpiration rate of the plants (g/s).

3.2. Greenhouse thermal model.

Temperature and relative humidity are commonly measured air properties, highly coupled through non-linear thermodynamic laws; for example

$$w = f(T, RH, P) = \frac{0.62198.P_{ws}(T).RH}{P - P_{ws}(T).RH} \tag{7}$$

where P is atmospheric pressure (kPa) and P_{ws} is saturation pressure of water vapor. Conversion of relative humidity to absolute water content, using the thermodynamic equation (7), provides a first step towards a state decoupled and linearized system.

We define a specific enthalpy change (H_s) as the energy per unit volume (Jm^{-3}) carried by the ventilating air. A thermal balance, neglecting enthalpy of incoming air and conductive heat losses from the greenhouse, yields

$$H_s \cdot \dot{V} = S_i \Rightarrow H_s = \frac{S_i}{\dot{V}} \tag{8}$$

The actuating capacity q_{fog}^{max} is selected to ensure that ventilation air changed ($\dot{V}^{max}$) can be saturated under any load conditions. Moreover, let w_{wet}^s, w_{fog}^s be the water carrying capacity of the saturated air for wet-pad or fog system operation respectively, and q_{wet}^s, q_{fog}^s be the effective water carrying capacity, from w_{out} to saturation, for wet-pad and fog systems respectively (see Figure 1). The actuating limit is $q_{fog}^{max} = q_{fog}^s \dot{V}$.

Maximum cooling is achieved when maximum evaporated water is used for a given ventilation rate. Then a controls capacity or controls feasible region is defined based on maximum ventilation capacity. In this condition the minimum specific enthalpy is:

$$H_s^{min} = \frac{1}{\dot{V}^{max}} S_i \tag{9}$$

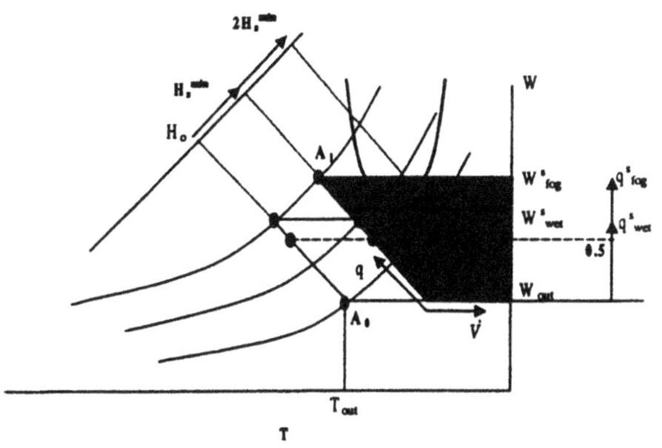

Fig. 1. Actuation limits defined by psychrometric properties

Equation (8) defines the feasible regime to the right of line A1A2, drawn as the locus of $H_o + H_s^{min}$, as shown in Figure 1. For example at half capacity, for $q = q_{fog}^{max}/2$ and $\dot{V} = \dot{V}^{max}/2$, that is $H_s = 2H_s^{min}$, starting from outside conditions at point <A$_0$>, the operating point will be <A$_3$> instead of <A$_1$> at full capacity.

The decision for a desired point of operation inside the feasible region is based on a cost function of the form

$$J = c_1(T_{in,sp} - T_{in,d}) + c_2(RH_{in,sp} - RH_{in,d}) + c_3\dot{V} + c_4 q_{fog}$$

Depending on the outside air conditions and the load S_i, the achievable conditions, for any cost, may not be the desirable ones ($T_{in,d}$, RH_{ind}). A rule base can be used to assign values for cost parameters c_1 and c_2 so as to equalize the risk on the crop from the deviations ($T_{in,sp}$-$T_{in,d}$) and ($RH_{in,sp}$-$RH_{in,d}$). In an attempt to use complete functionals for cost calculations, without resorting to fuzzy rules for cost assignments, we used the following extended cost function

$$J = c_1(T_{in,sp} - T_{in,d}) + \frac{c_1^*}{T_{in,sp} - T_{in,max}} + c_2(RH_{in,sp} - RH_{in,d}) + \frac{c_2^*}{1 - RH_{in,sp}} + c_3\dot{V} + c_4 q_{fog} \tag{10}$$

The added new terms are weighted such that the calculated set-points for temperature and humidity are kept away from an absolute maximum temperature (chosen by intuition and constraints for crop safety) and from the saturation line (risk of disease).

Using equations (7)-(9), the load $Env(S,T_o,RH_o)$ and a gradient descent method to minimize (10) the precompensator and variable translator (PVT) of Figure 2, calculates the realizable desirable target conditions $T_{in,sp}$ and $w_{in,sp}$, as well as the control values of q_{fog} and $\dot{V}$, which can be used as feedforward values, and other variables useful for the calculations at the controller. The PVT has all the

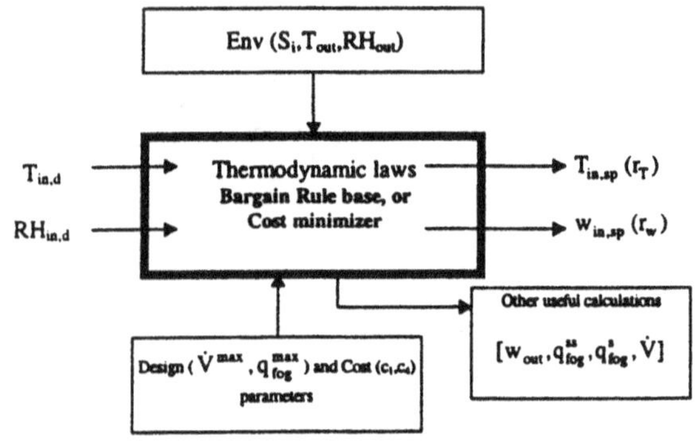

Fig. 2. Precompensator and Variable Translator calculating feasible control targets

required logic to compute realizable set-points and avoid pitfalls (i.e. singular values in Δ calculations of equation (13) given below) by post-processing the solution arrived by equation (10).

3.3. *Control model.*

For summer operation, q_{heater} in equation (6a) is set to zero. It is also worth noticing that, in a first approximation, the evapotranspiration rate $E(S_i(t), w_{in}(t))$, is in most part related to the intercepted solar radiant energy, through the following simplified relation

$$E(S_i(t), w_{in}(t)) = \alpha\frac{S_i(t)}{\lambda} - \beta_T w_{in}(t) \qquad (11)$$

where α is an overall coefficient to account for shading and leaf area index, and β_T is an overall coefficient to account for thermodynamic constants and other factors affecting evapotranspiration (i.e. stomata, air motion, etc).

On the basis of these observations, relations (6a) and (6b) take the forms

$$\frac{dT_{in}(t)}{dt} = \frac{1}{\rho C_p V}[S_i(t) - \lambda q_{fog}(t)] - \frac{\dot{V}}{V}[T_{in}(t) - T_{out}(t)] - \frac{UA}{\rho C_p V}[T_{in}(t) - T_{out}(t)] \qquad (12a)$$

$$\frac{dw_{in}(t)}{dt} = -\frac{\beta_T}{V}w_{in}(t) + \frac{1}{V}q_{fog}(t) + \frac{\alpha}{\lambda V}S_i(t) - \frac{\dot{V}}{V}[w_{in}(t) - w_{out}(t)] \qquad (12b)$$

Equations (12) are coupled nonlinear equations, which cannot be put into the rather familiar form of an affine analytic nonlinear system, due to their complexity appearing as the cross-product terms between control and disturbance variables. However, relations (12) can alternatively be written in the form (2), where, in the present case

$$\mathbf{x} = [x_1 \quad x_2]^T \triangleq [T_{in} \quad w_{in}]^T \ , \ \mathbf{y} = \mathbf{x}$$

$$\mathbf{u} = [u_1 \quad u_2]^T \triangleq [\dot{V} \quad q_{fog}]^T$$

$$\mathbf{v} = [v_1 \quad v_2 \quad v_3]^T \triangleq [S_i \quad T_{out} \quad w_{out}]^T$$

$$r_1 = r_2 = 1$$

$$f_1(\mathbf{x},\mathbf{v}) = -\frac{UA}{\rho C_p V}x_1(t) + \frac{1}{\rho C_p V}v_1(t) + \frac{UA}{\rho C_p V}v_2(t)$$

$$f_2(\mathbf{x},\mathbf{v}) = -\frac{\beta_T}{V}x_2(t) + \frac{\alpha}{\lambda V}v_1(t)$$

$$\mathbf{g}_1^T(\mathbf{x},\mathbf{v}) = \left[\frac{1}{V}(v_2(t) - x_1(t)) \quad -\frac{\lambda}{\rho C_p V}\right]$$

$$\mathbf{g}_1^T(\mathbf{x},\mathbf{v}) = \left[\frac{1}{V}(v_3(t) - x_2(t)) \quad \frac{1}{V}\right]$$

Note that disturbance variables of the greenhouse heating-cooling ventilating model, can be easily measured by the instrumentation installed in the greenhouse meteorological cage. Furthermore, the complexity of such systems is rather eased by the fact that the system state changes slowly and some state dependent parameters (i.e. β_T) can be considered constant (i.e. quasi-static system operation). Therefore, in the present case, a combined scheme of feedback with simultaneous feedforward linearization is plausible.

4. CONTROL OF THE GREENHOUSE VENTILATION MODEL.

In this section, the control method presented in Section 2, is applied to the above greenhouse ventilation model. To this end, in the present case, matrix $\mathbf{D(x,v)}$ is given by

$$D(x,v)=\frac{1}{V}\begin{bmatrix} v_2(t)-x_1(t) & -\frac{\lambda}{\rho C_p} \\ v_3(t)-x_2(t) & 1 \end{bmatrix}$$

whose determinant $\Delta(t)$ is given by

$$\Delta(t)=\frac{1}{V^2}\left[v_2(t)-x_1(t)+\frac{\lambda}{\rho C_p}(v_3(t)-x_2(t))\right] \quad (13)$$

which must be nonzero, for the system to be I/O linearized, decoupled and disturbance isolated. Note that, in the present case, the sum of the relative degrees equals system dimension, so there is no internal or zero dynamics. Note also that, in the case where, $\Delta(t)=0$, the input $u_1(t)$ affects the system states $x_1(t)$ and $x_2(t)$, with exactly the same way as $u_2(t)$, and thereby decoupling, as well as feedback-feedforward linearization is impossible.

By applying the control law of the form (3) the closed-loop system takes on the form

$$y_i^{(1)}=\hat{u}_i \quad , \ i=1,2$$

Moreover, in order to fix the dynamics of the output y_i, we apply the outer control laws of the form

$$\hat{u}_i=-a_{i0}y_i+b_i\tilde{u}_i \triangleq -\frac{1}{\tau_i}(y_i-\tilde{u}_i) \quad , \ i=1,2$$

and we obtain

$$y_i^{(1)}+\frac{1}{\tau_i}y_i=\frac{1}{\tau_i}\tilde{u}_i \quad , \ i=1,2$$

or in transfer function form

$$H_i(s)=\frac{1}{\tau_i s+1} \quad , \ i=1,2 \quad (14)$$

where, τ_i, i=1,2, are the time constants of the new closed-loop systems.

The above control algorithm can be summarized in the following two relations

$$u_1(t)=Q^{-1}(t)\left[\frac{\rho C_p}{\tau_1}\tilde{u}_1(t)+\frac{\lambda V}{\tau_2}\tilde{u}_2(t)-(\alpha+1)v_1(t)-UAv_2(t)\right.$$
$$\left.\left(UA-\frac{V\rho C_p}{\tau_1}\right)x_1+\left(\beta_T\lambda-\frac{V\lambda}{\tau_2}\right)x_2\right]$$

$$u_2(t)=\frac{\left(\left(UA-\frac{V\rho C_p}{\tau_1}\right)x_1+\frac{V\rho C_p}{\tau_1}\tilde{u}_1-v_1-UAv_2\right)(x_2(t)-v_3(t))}{Q(t)}$$
$$+\frac{\rho C_p(-x_1(t)+v_2(t))\left(\left(\beta_T-\frac{V}{\tau_2}\right)x_2+\frac{V}{\tau_2}\tilde{u}_2-\frac{\alpha}{\lambda}v_1\right)}{Q(t)}$$

where

$$Q(t)=\rho C_p[v_2(t)-x_1(t)]+\lambda[v_3(t)-x_2(t)]$$

and is depicted in Figure 3.

The greenhouse interior temperature and relative humidity are measured by a thermometer and a hygrometer, respectively, which usually are located a certain distance from the greenhouse ventilators and the fog or wet-pad system. Hygrometers also present a lag time themselves. Hence, the changes in the temperature and absolute humidity are determined after a certain time delay. Moreover, transport delays as well as unmodelled dynamics contribute to additional time lags. Therefore, an overall dead time, d_1 and d_2, must be considered for each output, y_1 and y_2, respectively. However, one must keep in mind that the nonlinear feedback-feedforward control law (9.9a), (9.9b), which renders the overall system linear and decoupled, relies on current state and disturbance measurements. Therefore, time delays may affect the feedback-feedforward linearization procedure and could degrade its performance. In order to avoid this problem, one must select τ_1 and τ_2, which are related to the speed of the closed-loop system response, to be large enough, resulting to a relatively slow closed-loop system. For example, a choice of $\tau_1>4d_1$ and $\tau_2>4d_2$ appears to be quite satisfactory compromise between the speed of the closed-loop system response and the performance of the feedback-feedforward linearizing control law. However, when faster responses are desired, then in order to avoid problems interwoven with the performance of the feedback-feedforward linearization procedure, one must utilize a Smith predictor, which, in addition, can compensate for large times delays d_1 or d_2.

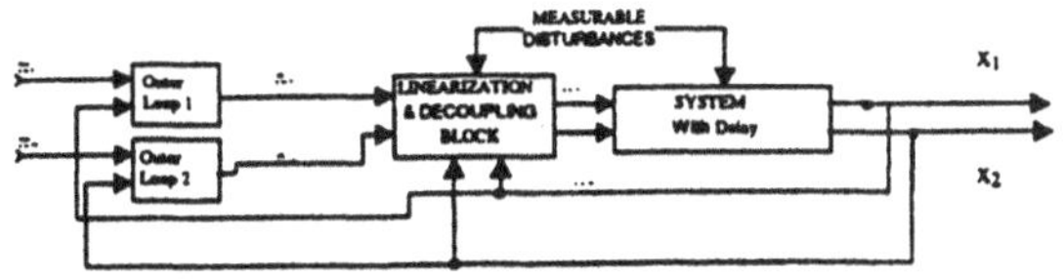

Fig. 3. Overall control strategy in case of small time delays and/or a slow desired response.

As it will be shown in the next Section, the proposed control algorithm, based on feedback/feedforward linearization and outer loop controllers, is quite robust to system parametric uncertainty as well as load disturbances. In particular, a 10% ubcertainty can be easily tolerated by the proposed controller. However, in the case of large parameter variations (e.g. plant growth that affects the greenhouse thermal capacity as well as evapotranspiration), one must apply more sophisticated control algorithms (like robust control or adaptive control algorithms) in order to compensate for such variations. Research on these topics (e.g. along the lines reported in Sigrimis *et al.*, 1999; Arvanitis *et al. 2000*), is cur-rently in progress.

5. SIMULATION RESULTS.

In this Section, the effectiveness of the proposed control scheme will be demonstrated by a case study. In particular, we consider here a greenhouse having an area of 1000 m^2 and a height of 4 m. The greenhouse is equipped by a shading screen, which reduces the incident solar radiant energy by 60%. The maximum water capacity of the fog system is 26 $g/min/m^3$. Maximum ventilation rate corresponds to 20 alternations of the greenhouse air per hour. Parameter α/λ takes the value 3.32×10^{-3} g/min/W, while β_T is negligible. The heat transfer coefficient is 25 $kW.K^{-1}$. Finally, we consider that unmodelled system as well as sensor dynamics contributes an overall dead time of 0.5 min in both temperature and humidity measurements. That is $d_1 = d_2 = 0.5\,\text{min}$.

A first simulation experiment has been conducted, in order to demonstrate the ability of the proposed control scheme to provide non-interacting control, and smooth closed-loop response to set-point step changes. To this end, we select the time constants of the two closed-loop subsystems to be $\tau_1 = \tau_2 = 5$ min. Then, after applying the feedback plus feed-forward linearizing and decoupling control law, we obtain the decoupled systems of the form (14). With this controller, we obtain the responses depicted in Figures 4 and 5. These figures illustrate the response for a set-point step change of absolute humidity from 18 gr/m^3 to 24 gr/m^3 (which corresponds to a relative humidity change from 60% to 80%), at t=100 min, while the temperature set-point remains at 30 °C, and the response for a temperature set-point step change from 30 °C to 28 °C, at t=200 min, while the absolute humidity set-point remains at 24 gr/m^3. Figure 6 illustrates the normalized (with respect to their maximum values) controller outputs. Note that in performing the simulation, the outside weather conditions are assumed to be T_{out}=35°C and w_{out}=4gr/m^3 (RH=10%), while S_i=300$Watt/m^2$. From the simulation results, it is clear that non-interacting

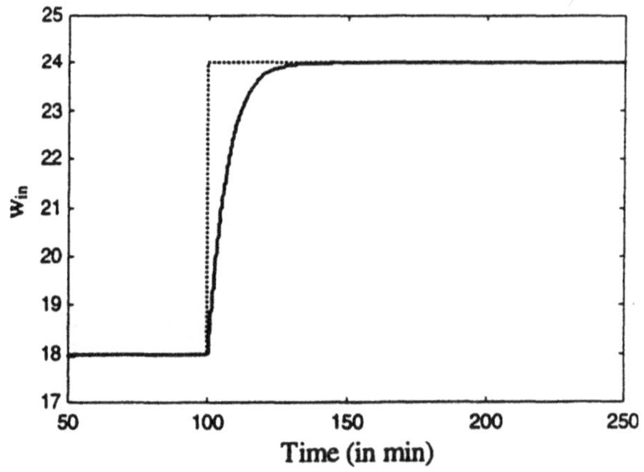

Fig. 4. Response of absolute humidity w_{in} for step changes in both humidity and temperature

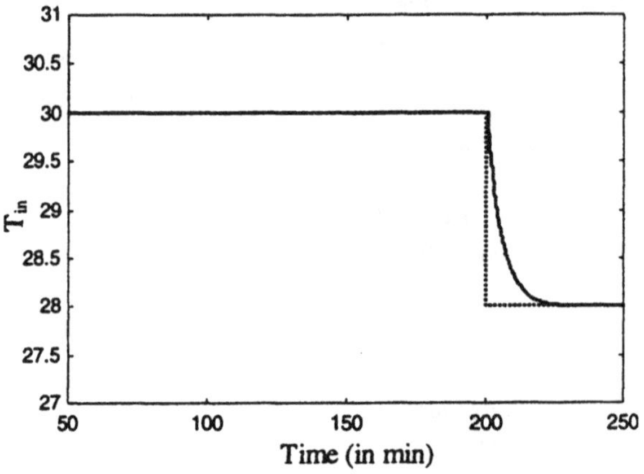

Fig. 5. Response of temperature T_{in} for step changes in both humidity and temperature

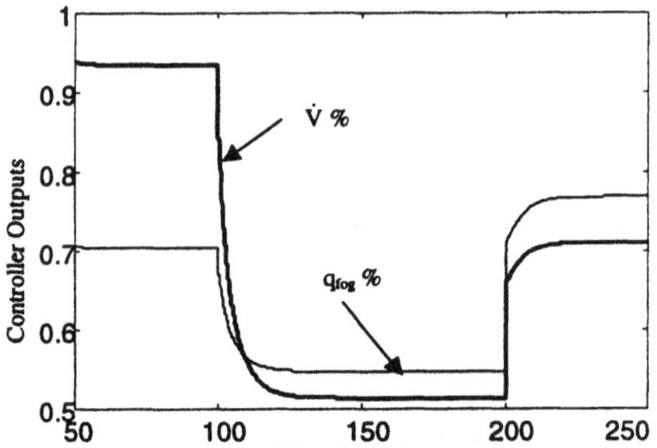

Fig. 6. Controller outputs for step changes in both humidity and temperature

control has been attained, while the closed-loop system response is the desired one.

The purpose of a second simulation experiment is to demonstrate that closed-loop system response is not affected by weather conditions, as it is expected, since the feedforward term of the linearizing/ decoupling controller compensate for system exter-nal disturbances. Here, the desired set-points are

$T_{in,sp}$=30 °C and $w_{in,sp}$=18 g/m^3. In order to perform the simulation, step changes in S_i, from 250 to 300 W/m^2, in T_{out}, from 35 °C to 32 °C and in w_{out}, from 4 to 8 g/m^3, have been applied, at time instants t=100 min, 150 min and 200 min, respectively. In the case where, there is no uncertainty in the model parameters, there is no effect of weather conditions on T_{in} and w_{in}. In this case, the controller outputs are depicted in Figure 7. In the case where a 10% uncertainty is considered in system parameters, then T_{in} and w_{in} are affected by weather conditions. However, two additional PID controllers (with the same controller parameters, K_P=0.5, K_I=0.25, K_D=0.25) are used. These PID controllers are able to provide fast regulatory control, as shown in Figures 8 and 9. Finally, in this case, the controller outputs are depicted in Figure 10. From this simulation experiment it becomes clear that, although there is a considerable uncertainty, the system remains decoupled and well behaved.

Finally, a simulation study has been accomplished in order to perform simultaneous temperature and humidity control in a greenhouse, in case of real weather conditions. To this end, weather data from a full summer day (June 3, 1999) in Arizona, U.S.A.,

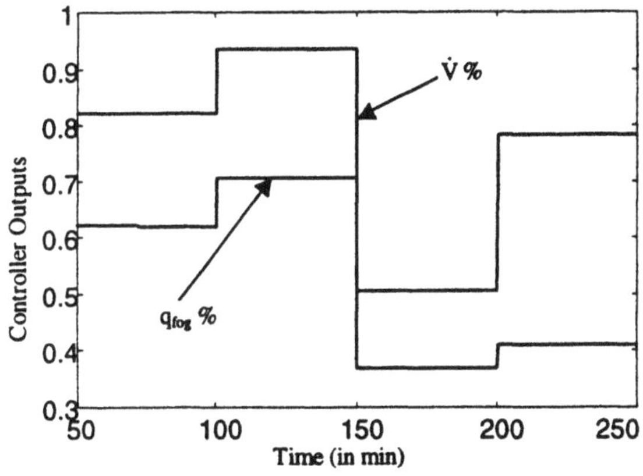

Fig. 7. Controller outputs for step changes in external disturbances (weather conditions)

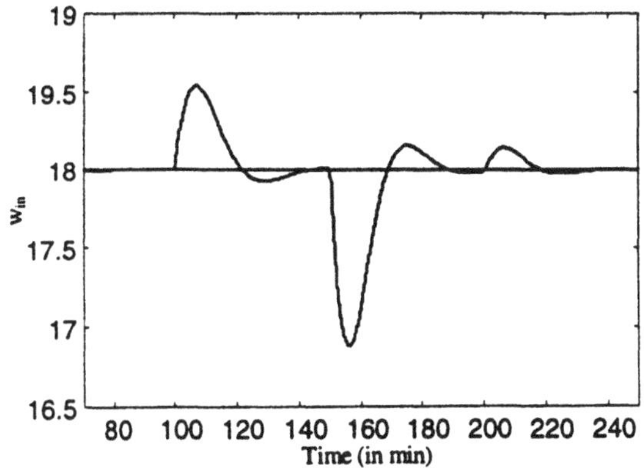

Fig. 8. Regulation of absolute humidity w_{in} for step changes in external disturbances in case of uncertainty

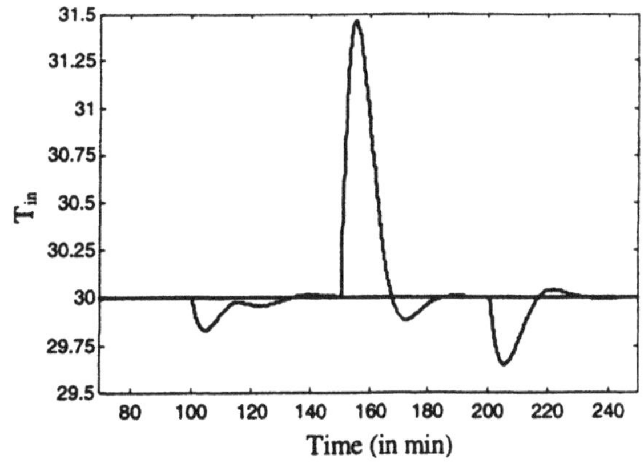

Fig. 9. Regulation of temperature T_{in} for step changes in external disturbances in case of uncertainty

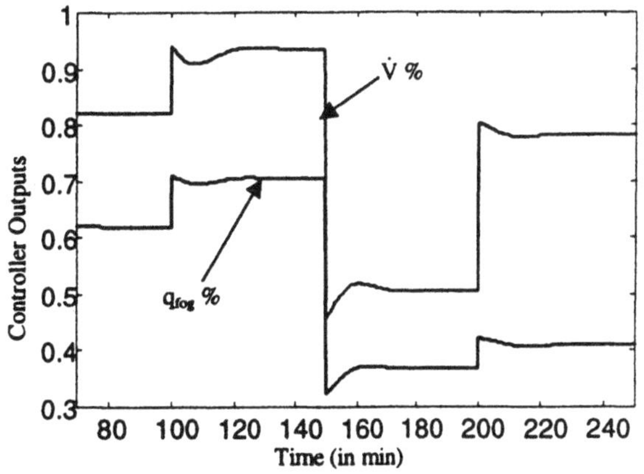

Fig. 10. Controller outputs for step changes in external disturbances in case of uncertainty

have been used. Set points for w_{in} and T_{in} have been obtained as outputs of the PVT block, and are illustrated in Figures 11 and 12, together with the trajectories of w_{in}, w_{out} and T_{in}, S_i, T_{out}, respectively. Obviously, the tracking performance of the proposed controller is remarkable. Finally, Figure 13 illustrates the controller outputs for this case.

6. CONCLUSIONS

The presented method of decoupling a highly non-linear and coupled system proved to be very effective in meeting formal requirements for control such as set-point tracking and disturbance rejection. The precompensator block to compute actuation limits and gains and variable change, using air psychrometric properties is a powerful approach to enable decoupling and linearization around the operating point. The method is currently implemented in MACQU (Sigrimis et al, 2000) systems to be placed in field operation.

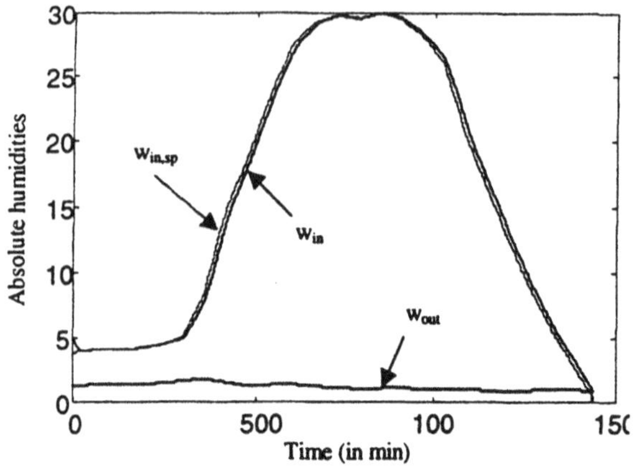

Fig. 11. Absolute humidity trajectories in case of simultaneous absolute humidity and temperature tracking

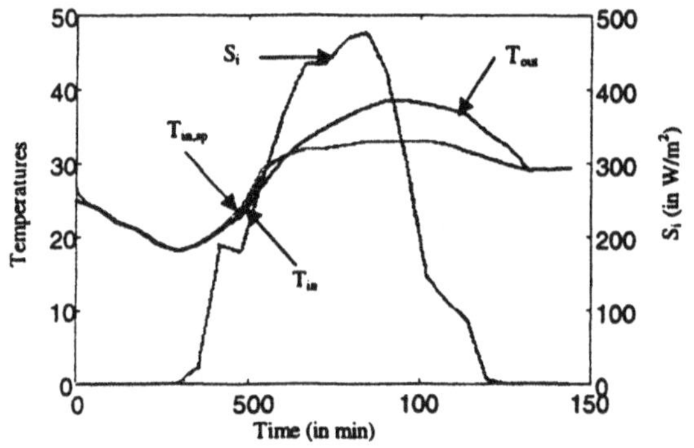

Fig. 12. Temperature trajectories in case of simultaneous absolute humidity and temperature tracking

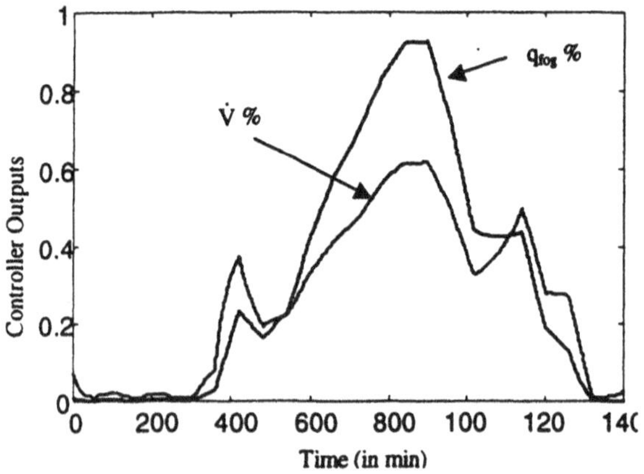

Fig. 13. Controller outputs in case of simultaneous absolute humidity and temperature tracking

Acknowledgement
This work is supported by the HORTIMED (ICA3-CT1999-00009) project to enable collaborative management of the root and shoot greenhouse management. MACQUD project (EU-DGVI PL98-4310) provides the MACQU technology for easy field implementation.

REFERENCES

Arvanitis, K.G., P.N.Paraskevopoulos and A.A.Vernardos, 2000. Multirate adaptive temperature control of greebhouses. *Comp. Electronics Agric.*, **26**, 303-320.

Chao, K., R.S. Gates and H.-C. Chi, 1995. Diagnostic hardware/software system for environment controllers. *Trans. ASAE*, **38**, 939-947.

Chao, K. and R.S. Gates, 1996. Design of switching control systems for ventilated greenhouses. *Trans. ASAE*, **39**, 1513-1523.

Chao, K., R.S. Gates and N. Sigrimis, 2000. "Fuzzy logic controller design for staged heating and ventilating systems. *Trans. ASAE*, **43**, 1885-1894.

Gates, R.S. and D.G. Overhults, 1991. Field evaluation of integrated environmental controllers. *ASAE Paper No. 91-4037*, St.Joseph, Mich.: ASAE.

Jones, P., J.W.Jones, L.H.Allen Jr. and J.W.Mishoe, 1984. Dynamic computer control of closed environmental plant chambers: Design and verification. *Trans. ASAE*, **27**, 879-888.

Sigrimis, N., K.G.Arvanitis, I.K.Kookos and P.N.Paraskevopoulos, 1999. H^∞-PI controller tuning for greenhouse temperature control, *Proc. 14th IFAC Triennial World Congr.*, **K**, 485-490, Beijing, China, July 5-9, 1999.

Sigrimis N, K.G. Arvanitis, G. Pasgianos, A. Anastasiou, K.P. Ferentinos, 2000. NEW WAYS ON SUPERVISORY CONTROL: a virtual greenhouse: to train, to control and to manage. IFAC –MIM conference, July 14-16 Patras Rio Greece.

Stanghellini, C., van Meuss, W.T.M., 1992. Environmental control of greenhouse crop transpiration. *J.Agric. Eng. Res.*, **51**, 297-311.

Stanghellini, C., De Jong, T., 1995. A model of humidity and its applications in a greenhouse. *Agric. For. Meteorol.*, 76, 129-148.

Zhang, Y. and E.M.Barber, 1993. Variable ventilation rate control below the heat-deficit temperature in cold-climate livestock buildings. *Trans. ASAE*, **36**, 1473-1482.

Zolnier, S., R.S.Gates, J.Buxton and C.Mach, 2000. Psychrometric and ventilation constraints for vapor pressure deficit control. *Comp. Electronics Agric.*, **26**, 343-359.

MODELLING OF THE TEMPERATURE UNDER GREENHOUSE BY MULTI-MODEL APPROACH

E. Feki, R. Mhiri, M. Ksouri

Ecole Nationale des Ingénieurs de Tunis, B.P. 37,Belvèdère 1002 Tunis Tunisie
Tél: 2161874700, Fax: 2161510729 Email: elfeki@edunet.tn – Radhi.Mhiri@isetr.rnu.tn

Abstract: The methodology proposed in the paper applies multi-model approach to the modelling of same climate variables within a greenhouse. The nonlinear physical phenomena governing the dynamics of temperature and humidity in such systems are, in fact, difficult to model and control using traditional techniques. In this paper we will demonstrate the use of the multi-model approach in the modelling of the temperature under greenhouse. We will show the interest of this approach in comparison with classical approach. *Copyright 2001© IFAC*

Keywords: greenhouse, modelling, temperature, multi-model.

1. INTRODUCTION

The greenhouse situated in the Mediterranean regions require the control of the climate, so that to maintain the agriculture in good conditions which are compatible with the agronomic and economic objectives of the farmer. Indeed the control and study of the evolution of the various technological parameters which condition the environment of the agriculture (temperature, humidity and sunlight) require a control in real time. Thus the recourse to the computerised proves to be essential. We turn around the micro-computer to manage the climatic conduct under greenhouse.

This paper shows how an integrated computer system can be used to improve the climatic performance of a greenhouse by controlling its temperature and humidity. Furthermore the paper give some results in order to demonstrate the validity of the use modelling by multi-model approach of greenhouse climate. Finally, the comparison between this approach and a classical technique in a modelling problem will show the interest of the proposed approach.

2. MATERIEL AND METHODS

This application concerns an experimental greenhouse at the I.N.R.S.T. located in the Borj-Essedria Area, a suburb 20Km south of Tunis. The climate of the greenhouse is controlled by a cooling-heating system, a ventilation system, a moistening system and an artificial lighting system.

Fig. 1. General view of the interior of the experimental greenhouse

2.1 Description of the control system

The greenhouse is used for vine plants which require a temperature around 24°C and a humidity between 60% and 70%. The experimentation consists of placing a station of acquisition and climatic control. It involves a group of hardware and software which permit the acquisition and numerical storage of the climatic data such as : temperature, humidity, and radiation (Feki, et al., 1996). The control system is based around a personnel computer. The acquisition of data is supplied by the use of a card of analogical/digital conversion (12bits) and the commands will be achieved though a command's card (PCL724).

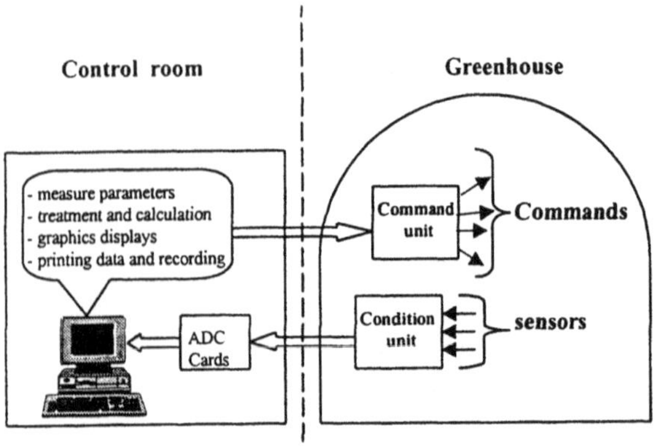

Fig. 2. The schematic diagram of the control system

The aim of the control is to ensure a temperature around 24°C during the day and around 20°C during the night. The humidity should be around 70%. When the sunlight falls under 5000 lux, the artificial lighting system must be set on.

2.2 Measurements with classical control

The management and piloting software of the greenhouse allows to fix two temperature thresholds, minimal and maximal ones.

When the greenhouse temperature reaches a minimum level, the heating-system is actuated. In contrast, the greenhouse stops heating when it exceeds the maximum level.

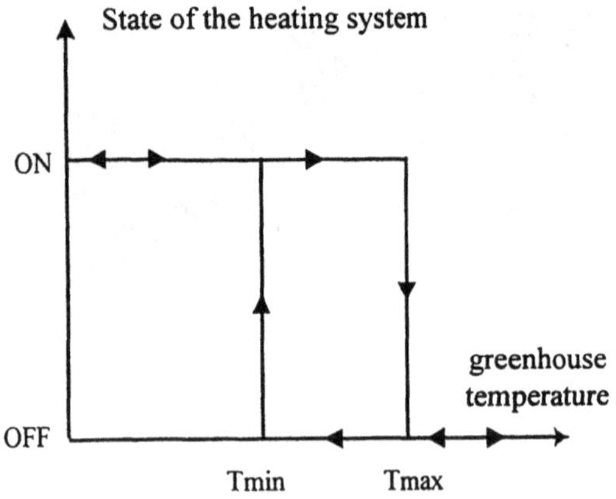

Fig. 3. Classical ON-OFF control

According to these recording we can conclude these results show that the temperature remains within the desired range (between 20°C and 26°C) and that humidity is always over the 70% mark. These climatic conditions permitted to supply vigorous vineyards consequently to exploit the greenhouse throughout the year.

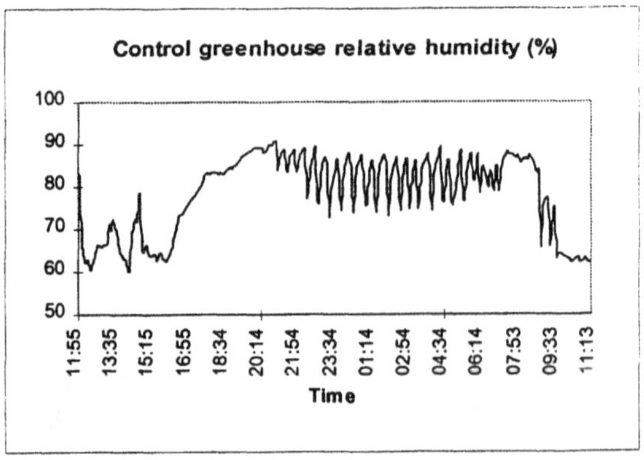

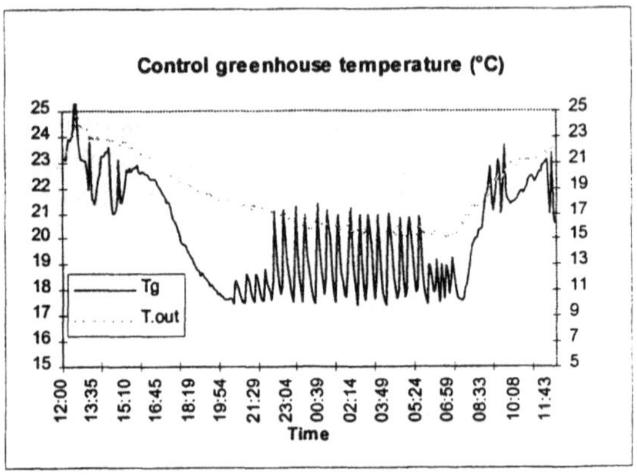

Fig. 4. Evolution of the temperature and the humidity under greenhouse with control during one day in October

3. PHYSICAL MODEL FOR GREENHOUSE

3.1 Introduction

To design efficient environmental controllers for greenhouses it is necessary to develop models that adequately describe the system to be controlled. These models must be related to the external influences of the outsides weather conditions (such as solar radiation, outside air temperature, wind speed, etc.), and to the actuators used in the greenhouse (such as ventilators, cooling systems, heating systems, among others) (Boaventura Cunha, et al., 1997).

Basically, there are two different methods for computing the models. One is based in terms of the physical laws involved in the process, and the other is based on an analysis of the input-output data of the process. In the first method the thermodynamic properties of the greenhouse system are employed. The greenhouse thermal performance could be described, for instance, by the use of a single energy balance equation (Boulard, et al., 1993).

$$\eta G_0 - K_S \Delta T - K_l \Delta e - K_c \Delta T + Q_h - Q_m = 0 \ (Wm^{-2}) \quad (1)$$

Where η is the solar efficiency, K_c ($Wm^{-2}K^{-1}$), K_s($Wm^{-2}K^{-1}$), K_l($Wm^{-2}Pa^{-1}$) are the coefficients of the heat transfer for the cover material, the ventilation heat exchange for sensible heat and the ventilation heat exchange for latent heat, respectively, $Q_m(Wm^{-2})$ is the heat storage rate, $G_0(Wm^{-2})$ is the outside global radiation, $Q_h(Wm^{-2})$ is the heat provided by the heating system, $\Delta T(K)$ is the difference between the inside and outside air temperature, and $\Delta e(Pa)$ is the difference between the inside and outside water vapour pressures.

Despite the relatively simple form of equation, it must take into account the fact that the above parameters are time-variant and weather-dependent. As an example, the solar efficiency depends on the reflection, absorption and transmission properties of the transparent cover and the opaque components. Since the optical properties of the transparent materials and the angle of incoming radiation, among others, vary throughout the day and the growth season, the solar efficiency of the greenhouse has a varying value. Therefore, in most applications it is difficult to obtain a satisfactory mathematical model of the greenhouse thermal system from physical knowledge only (Udlink Ten cate, 1987).

The second method is based on experimentation where the input (u) and output (y) signals from the system to be identified (see fig. 5.) are recorded and subjected to data analysis in order to infer a model. In this figure *e* denotes a noise signal. Here, y' and ε represent the model output and the prediction error. This procedure is known as system identification.

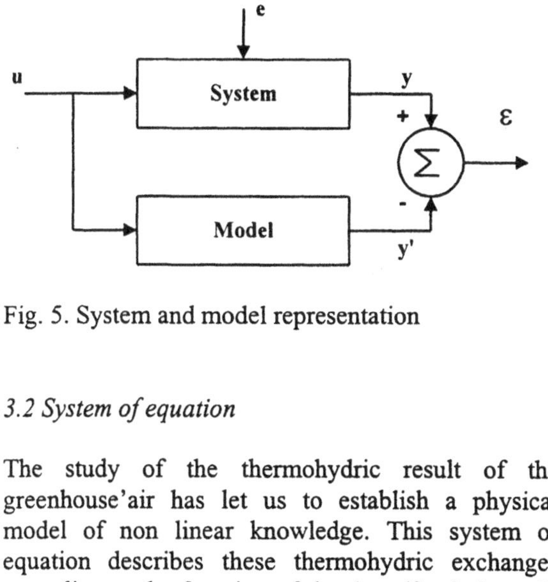

Fig. 5. System and model representation

3.2 System of equation

The study of the thermohydric result of the greenhouse'air has let us to establish a physical model of non linear knowledge. This system of equation describes these thermohydric exchanges according to the function of the time (Souissi, et al., 1996).

$$\frac{T_i(k+1)-T_i(k)}{T_e^*}=(\frac{1}{\rho_i V_i C_i})[K_s R_s+S_c(AR_s+B(W_i^*(T_i)-W_i)) +Q_c+\lambda_p S_p\frac{(T_e-T_i)}{e_p}+\rho_i R_a V_i C_i(T_e-T_i)+K_v V_e] \quad (2)$$

$$\frac{W_i(k+1)-W_i(k)}{T_e^*}=(\frac{1}{\rho_i V_i C_i})[(\frac{S_c}{L})(AR_s+B(W_i^*(T_i)-W_i)) +\rho_i R_a V_i(W_e-W_i)+F_r] \quad (3)$$

Notation: Symbols

A : Coefficient of radiation
B : Coefficient of convection
C_i : Specific heat of the air
Q_c : heat flux from heater
e_p : thickness of the cover
F_r : moistening flux
L : heat of evaporation of the water
R_a : Rate of renewal of the air
R_s : outside global radiation
S_c : area of the culture
S_p : area of the cover material
T_i : inside temperature
T_e : outside temperature
V_i : volume of the greenhouse
W_i : humidity absolved from the internal air of the greenhouse
W_e : humidity absolved from the external air of the greenhouse
$W_i^*(T_i)$: saturating absolved humidity in the temperature T_i
λ_p : specific conductivity of the cover
ρ_i : density of the air

3.3 Linearisation of the model

The linearisation of the model around a point of functioning ends in the following system of equations shelf space :

$$T_i(k+1)=a_{11}T_i(k)+a_{12}W_i(k)+b_{11}Q_c(k)+ b_{12}R_a(k)+b_{13}F_r(k)+d_{11}T_e(k)+d_{12}R_s(k)+ d_{13}W_e(k)+d_{14}V_e(k) \quad (4)$$

$$W_i(k+1)=a_{21}T_i(k)+a_{22}W_i(k)+b_{21}Q_c(k)+ b_{22}R_a(k)+b_{23}F_r(k)+d_{21}T_e(k)+d_{22}R_s(k) +d_{23}W_e(k) \quad (5)$$

The state variable of the greenhouse is:

- X_k: state variable, greenhouse air temperature and humidity.

$$X_k=\begin{bmatrix}T_{ik}\\W_{ik}\end{bmatrix}$$

- U_k: control variable

$$U_k=\begin{bmatrix}Q_{ck}\\Qf_k\\R_{ak}\\F_{rk}\end{bmatrix}$$

- V_k: disturbances, solar irradiation R_s, outside temperature T_e, outside humidity W_e, and speed of the wind V_v

$$V_k = \begin{bmatrix} T_{ek} \\ R_{sk} \\ w_{ek} \\ v_{vk} \end{bmatrix}$$

The climate inside the greenhouse can be represented by a linear equation :

$$\begin{aligned} X_{k+1} &= AX_k + BU_k + DV_k \\ Y_k &= CX_k \end{aligned} \qquad (6)$$

With

- A: state variable matrix

$$A_k = \begin{bmatrix} a_{11} & a_{12} \\ a_{21} & a_{22} \end{bmatrix}$$

- B: state command matrix

$$B_k = \begin{bmatrix} b_{11} & b_{12} & b_{13} & b_{14} \\ b_{21} & b_{22} & b_{23} & b_{24} \end{bmatrix}$$

D: state disturbances matrix

$$D_k = \begin{bmatrix} d_{11} & d_{12} & d_{13} & d_{14} \\ d_{21} & d_{22} & d_{23} & d_{24} \end{bmatrix}$$

In this work we are interested in the temperature under the greenhouse the first method of modelling consists of identifying a space of measure during 12 hours day only, one model which involves the three commands to be known : the heating system, ventilation system and the cooling system.

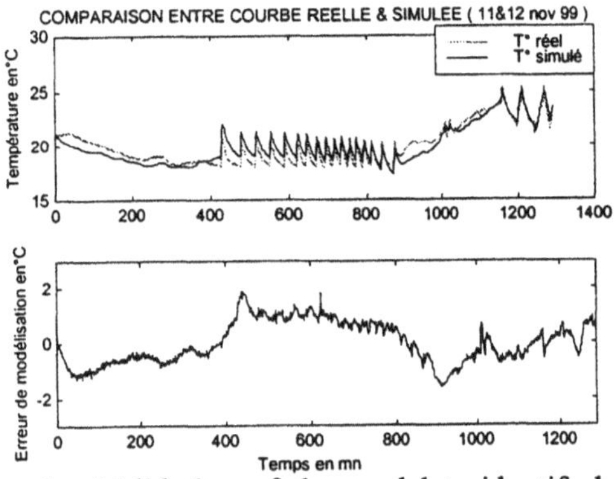

Fig. 6. : Validation of the model to identify by the classic approach of modelling

This figure shows that the results obtained by the algorithm are satisfying in whose measures are precise (for example the period when the heating system work), in contrast, in other spaces the error of modelling becomes important 2°C. For this reason we have recourse to the multi-model identification.

4. MULTI-MODEL APPROACH

The main objective in order to understand our daily life environment is therefore to find the simplest model which can describe efficiently the system we interest in.

But this system may also change in such away that it becomes impossible to use only one of its models. In the multimodel approach, it is defined a dynamic model M which interpolates all the different local ones M_i so that :

$$\begin{cases} M = f(M_i) \\ M = M_i \text{ if system } \in D_i \end{cases} \qquad (7)$$

Where D_i is the domain of validity of the model M_i, i.e. the domain where the model M_i is sufficient to represent the process, and f is a non-linear function (Dubois, et al., 1996).

If the domains of validity are separated (fig. 7.), then coefficients v_1 , v_2 can take only values 0 or 1, and verified the relation

$$v_1+v_2+...+v_N = 1 \qquad (8)$$

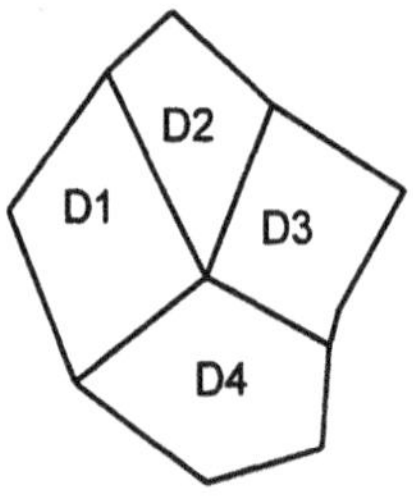

Fig.7. Separated domains of validity

Another situation is presented, If the domains of validity overlap fig. 8.

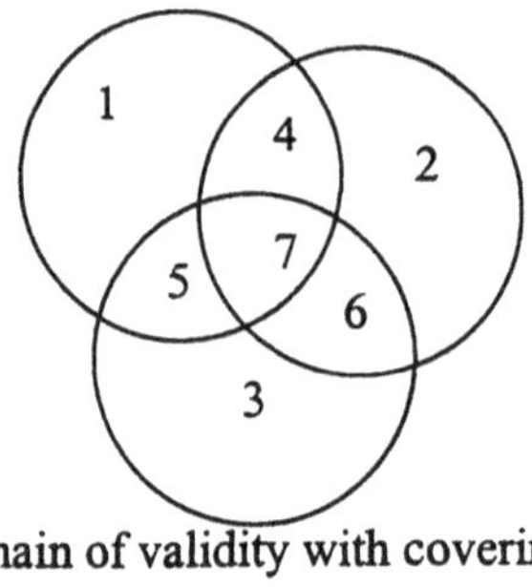

Fig. 8. Domain of validity with covering

In this case the state space x(t) is:

$$x(t) = v_1 x_1(t) + v_2 x_2(t) \qquad (9)$$

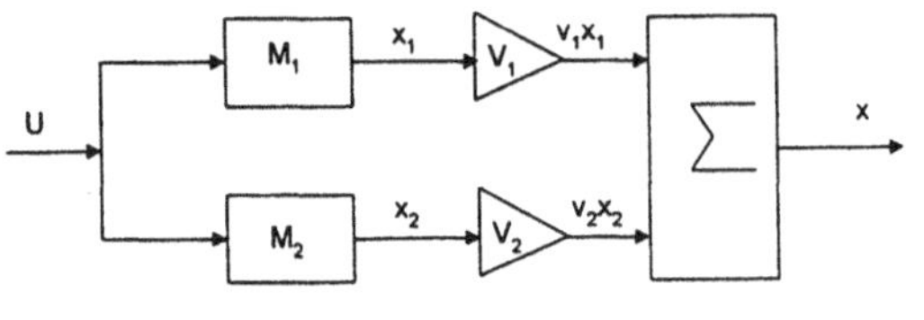

Fig. 9. Fusion of the state

4.1 Validity estimation using fuzzy logic

In this section, we propose the use of the fuzzy logic as a new approach to cope with the validity issue. Let us consider a process represented by the following non-linear model:

$$\begin{aligned} \dot{x}_i &= A_i x_i + B_i u_i \\ y_i &= C_i x_i \end{aligned} \tag{10}$$

It is possible to define models, considered to be the extreme models, in the base described by the equations:

$$\begin{aligned} \dot{x}(t) &= f(x,u,t) \\ y(t) &= g(x,u,t) \end{aligned} \tag{11}$$

The validity management through fuzzy logic is based on the following scheme:

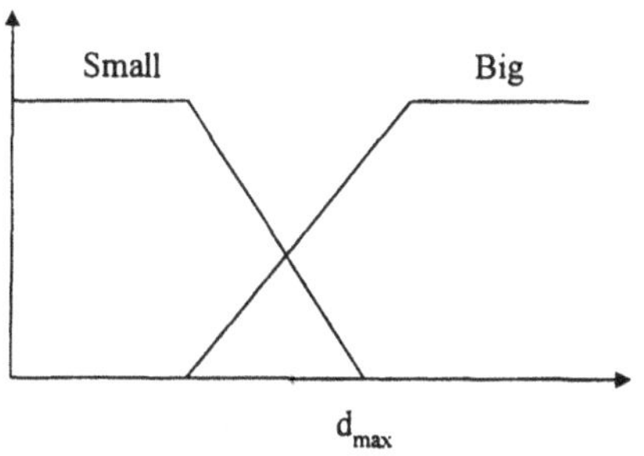

Fig. 10. Partition for the distance's universe of discourse

Fuzzy logic is mainly based on the used of qualitative terms as *small* or *big*. Therefore, it is first defined the relationships between the models and some arbitrary decomposition of the state space which is common for all the models (Dubois, et al., 1996).
Another way is to consider the distances d_i of the state vector to the domains of validity as the variable of the fuzzy model.
For instance, let us consider 2 fuzzy subsets small and big to describe this distance

The fuzzy rules base is given in :
IF d_1 is Small THEN Apply M_1
IF d_2 is Small THEN Apply M_2
IF d_3 is Small THEN Apply M_3
IF d_1 is Big AND d_2 is Big AND d_3 is Big THEN Apply M_1, M_2 and M_3

4.2 use Multimodelling approach in the greenhouse climate

We divide the group of measures in many models Mi with i=1 to 6. Each model is in fact a simplified mathematics formulation of the system. Simplified because the main objective of a multi-model approach is to allow the reduction of the complexity of the system while studying it under particular circumstances.

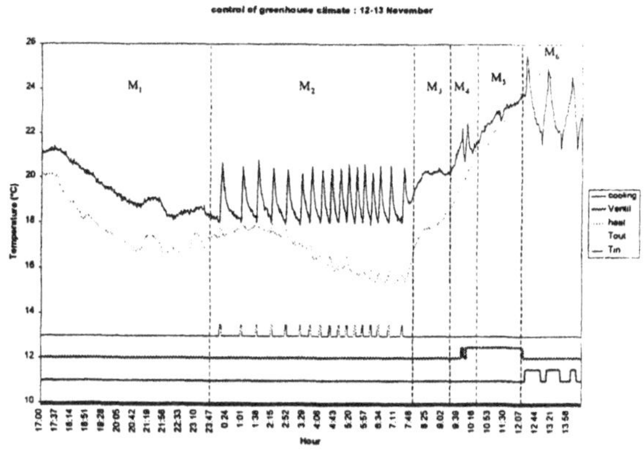

Fig. 11. Division of the group of measure of six models.

The choice of the number of the models and evolution of the system is represented by 6 different dynamism.
- **M1**: phase where the temperature under greenhouse goes down without reacting on any action by influence of the external elements: reduction in the outside temperature and the solar radiation, it's when the sun goes down.
- **M2**: phase where the heating system takes over thus we notice the ON/OFF of this system, it's the diurnal period.
- **M3**: phase on which one notices that the interior temperature increases under the sunrise: it's the morning (when the sun goes up).
- **M4**: model where the command ON/OFF of the fan system.
- **M5**: Phase where the fan system is ON but the temperature under the greenhouse continuous to increase
- **M6**: intervention of the cooling system ON/OFF.

To obtain a small transition between the models we use this expression:

$$M = f(M_i) = \sum_i a_i M_i \tag{12}$$

This gains are generated by the fuzzy logic, for example, if we have a transition between M1 and M2, so we define fuzzy subsets characterised by their membership functions according to the domains of validity.

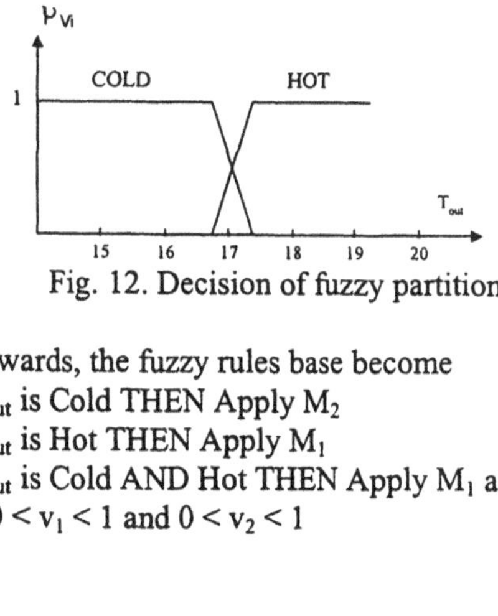

Fig. 12. Decision of fuzzy partition

Afterwards, the fuzzy rules base become
IF T_{out} is Cold THEN Apply M_2
IF T_{out} is Hot THEN Apply M_1
IF T_{out} is Cold AND Hot THEN Apply M_1 and M_2
So: $0 < v_1 < 1$ and $0 < v_2 < 1$

Results and discussion

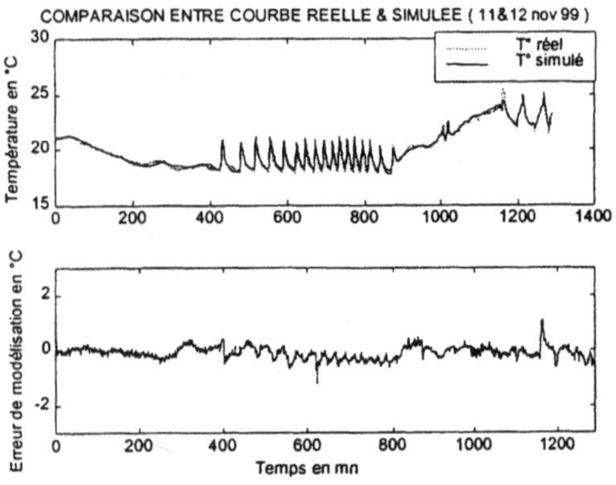

Fig. 13. Validation of the modelling by the multi-models approach

According to the results obtained by the multi-model modelling we point out that this approach is sufficient and in no case does the relative error reach 1°C.

5. CONCLUSION

In this article we used a multi-model approach with fusion of the models to model a complex system such as the greenhouse. We showed that one can overcome the difficulty of the variation of the parameters of the greenhouse by using multi-modelling to have valid models around their points of operation. The prospects is to apply multi-model approach to control greenhouse, and really bring us to implement this type of the commands on all the equipment of regulations of the greenhouse.

REFERENCES

Boaventura Cunha, J., C. Couto and A. E. Ruano (1997). Real-time parameter estimation of dynamic temperature models for greenhouse environmental control, *Control Eng. Practice*, Vol5 No10, pp.1473-1481.

Boulard, T. and A. Baille (1993). A simple greenhouse climate control model incorporating effects on ventilation and evaporative cooling, *Agricultural and forest meteorology 65*, pp145-157.

Delmotte, F., L. Dubois and P. Borne (1996). A genera scheme for multi-model controller using trust, *Mathematics and computers in simulation*, pp.173-186.

Dubois, L., T. Fukuda, F. Delmotte and P. Borne (1996). Multi-model systems are universal approximators *CESA'96*, Lille-France.

Feki, E., R. Mhiri, .M. Annabi, A. Ghorbel (1998). Control of greenhouse climate, *Proceeding of the CESA'98, Computational engineering in systems applications*, pp663-666.Tunisia.

Ksouri-Lahmari, M., A. El Kamel, M. Benrejeb and P. Borne (1997). Multimodel multicontrol decision making in system automation, *IEEE-SMC'97*, Orlando, Florida, USA.

Murray-Smith, R. and T. A. Johansen (1997), Multiple model approaches to modeling and control, Ed. *Taylor & Francis*.

Souissi, M., L. Sbita and M. Annabi (1996). from complete to a simple model greenhouse climate control, *World renewable energy congress VI.*, USA.

Udlink Ten cate, A. J. (1987). Analysis and synthesis of greenhouse climate controllers, *Computer applications in agricultural environments*, pp1-19, Butterworths, London.

OPTIMAL CONTROL OF NITRATE IN LETTUCE BY GRADIENT AND DIFFERENTIAL EVOLUTION ALGORITHMS

I.L. Lopez Cruz, L. G. Van Willigenburg, G. Van Straten

*Systems and Control Group, Wageningen University,
Bomenweg 4, 6703 HD, Wageningen, The Netherlands,
E-mail:Irineo.Lopez@user.aenf.wag-ur.nl*

Abstract: Since high concentration levels of nitrate in lettuce are undesirable, its control is currently an important problem in the context of European Union regulations. Using a dynamic model that predicts the amount of nitrate at harvest time, an optimal control problem is formulated and solved through an enhanced classical gradient method and a Differential Evolution algorithm. This work shows that in order to avoid local minima an efficient evolutionary algorithm may be applied to solve optimal control problems or to provide a good initial guess for a classical method, which solves smooth continuous-time optimal control problems more accurately and efficiently. *Copyright © 2001 IFAC*

Keywords: Optimal control, Artificial Intelligence, Genetic Algorithms, Global optimization, Gradient methods

1. INTRODUCTION

High concentration levels of nitrate in lettuce crop and other leafy vegetables are undesirable because they have a negative effect on human health. Therefore, methods are sought to control the nitrate levels of a greenhouse lettuce crop. As a first step a model of lettuce growth, which predicts the amount of nitrate content at harvest time, has been proposed (Seginer et al., 1998). An optimal control problem has been formulated and some properties of its solution have been analyzed using a simplified lettuce model (Ioslovich and Seginer, 2000). Also the full two-state nonlinear lettuce model has been used to get a numerical solution to another optimal control problem that uses light and temperature as control inputs by means of a first order gradient algorithm (Stigter and Van Straten, 2000). The aim of this paper is to solve a new optimal control problem that includes light, temperature and also carbon dioxide concentration as control inputs, by an evolutionary algorithm and to compare the results with those obtained by the Adjustable Control-variation Weight (ACW) gradient algorithm (Bryson, 1999, Weinreb, 1985). The evolutionary algorithm selected is the recently proposed Differential Evolution (DE) algorithm (Storn and Price, 1997), since DE is an evolutionary algorithm that can approximate the global optimum and is also very efficient computationally compared to other evolutionary algorithms. The paper is organized as follows: first a brief description of the optimal control problem is given; especially some properties of the dynamic lettuce model are emphasized. Then, the main characteristics of the Differential Evolution algorithm are outlined. Next, the results are described, compared and discussed.

2. OPTIMAL CONTROL OF NITRATE IN LETTUCE

The lettuce model is based on carbon balances of the vacuoles and the structure that prevail in the plant cells. The so-called NICOLET model (Seginer et al., 1998) has two state variables: carbon content in the vacuoles (M_{Cv} mol[C] m^{-2} [ground]) and carbon content in the structure (M_{Cs} mol[C] m^{-2} [ground]) that represent a carbon source-sink relation in the plant driven by sunlight, temperature and carbon dioxide concentration. Photosynthesis and growth can proceed uninhibited as long as the non-structural carbon concentration in the vacuoles remains within certain limits. However, when affected by environmental conditions the non-structural carbon concentration approaches zero, growth will be reduced. In the model, this transition is implemented by introducing a smooth switching function which is one for non-inhibiting levels, but falls off to zero rapidly when the assimilate stock becomes empty. When carbon assimilates in the vacuoles are too high a similar switching function brings photosynthesis to a halt.

The core of the model is given by two differential equations:

$$\dot{M}_{Cv} = F_{Cav} - h_g F_{Cm} - F_{Cg} - F_{Cvs} \quad (1)$$

$$\dot{M}_{Cs} = F_{Cvs} - (1 - h_g) F_{Cm} \quad (2)$$

which represent the main carbon balances. The F's in equation (1) denote the rates of photosynthesis, maintenance, growth respiration and uninhibited growth, respectively, which are functions of the states and the inputs light (I [mol PAR m^{-2}s^{-1}]), carbon dioxide (C_{Ca} [mol m^{-3}]) and temperature (T °C). h_g denotes the inhibition function for growth.

The description of all model equations and parameter values is given in Yarkoni and McKenna, (2000). Here it is worthwhile to outline that in the version used here (NICOLET B3) two important modifications were made as compared to the original NICOLET model (Seginer et al., 1998). In NICOLET version B3 the inhibition functions for photosynthesis and growth were changed in such a way that they take a value of zero when vacuolar carbon concentrations reach the appropriate bound. And also, as seen in Eqns (1) and (2), the new model incorporates the depletion of structural matter to meet the requirements of maintenance respiration when the carbon content in the vacuoles is low. This situation may occur when the model is exposed to a long darkness period. It turned out that the original model predicts negative values for the carbon content in the vacuoles when it was used in the solution of an optimal control problem by the differential evolution algorithm with three control inputs. From the states of the lettuce model several outputs such as dry and fresh matter, sugars and nitrate concentration are calculated. The nitrate concentration follows from M_{Cv} by a negative algebraic correlation, which expresses the plants policy to maintain its turgor pressure.

One formulation of the optimal control of nitrate in lettuce is as follows. While minimizing the integral of light

$$J = \int_0^{t_f} I(t)dt \quad (3)$$

calculate the control trajectories of light, carbon dioxide and temperature such that a desired fresh head weight of lettuce (y_{dfm} [gr]) and a specified amount of nitrate (y_{dNO_3} [ppm]) are obtained at a specified harvest time (t_f) i.e.,

$$y_{fm}(t_f) = y_{dfm} \quad (4)$$

$$y_{NO_3}(t_f) = y_{dNO_3} \quad (5)$$

where $y_{fm}(t_f)$ and $y_{NO_3}(t_f)$ are the corresponding outputs of the model. The control inputs are bounded because it is apparent that the light intensity cannot be negative and the same is true for carbon dioxide. Also the temperature must lie within a domain tolerated by the lettuce crop

$$I_{\min} \le I(t) \le I_{\max},$$
$$C_{\min} \le C_{Ca}(t) \le C_{\max},$$
$$T_{\min} \le T(t) \le T_{\max} \text{ for } 0 \le t \le t_f \quad (6)$$

3. DIFFERENTIAL EVOLUTION

A numerical solution to the specified optimal control problem can be obtained by indirect methods of optimal control like a first order gradient method, but also by direct methods. In this work a direct method based on evolutionary algorithms is used to approximate the global optimum solution. In the direct method the control trajectory is parameterised as a sequence of piece-wise constant values, which have to be found by the optimisation. The Differential Evolution algorithm is a kind of evolutionary algorithm that has recently been proposed for the solution of static parameter optimisation problems (Storn and Price, 1997). This algorithm has several nice properties over other evolutionary algorithms because it is easy to understand and very efficient computationally. An outline of this algorithm is presented in figure 1. As in other evolutionary algorithms main operators in

DE are mutation, crossover and selection. Yet, in contrast to archetype genetic algorithms, they use a floating-point representation for the solutions. Also the main operator in DE is rather different than in other evolutionary algorithms. Similarly to Evolution Strategies here each chromosome is represented as a real parameter vector $\mathbf{a} = [a_1,...,a_q]$, and it is required at generation g a population containing μ individuals $\mathbf{a}_i; i = 1,...,\mu$. The essential feature of differential evolution algorithm rests on the mutation operator. A so-called mutant vector (v_i), is generated by adding the weighted difference between two or four selected population vectors to another vector:

Figure 1. *Differential Evolution algorithm*

```
Generate random solutions that cover the
given space.
Evaluate each solution.
g=1;
while (convergence is not reached)
   for i=1 to Population Size
         Apply differential mutation.
         Execute differential crossover.
         Clip the new solution if
         necessary.
         Evaluate the new solution.
         Apply differential selection.
   end
   g=g+1;
end
```

$$\mathbf{v}_i = \mathbf{a}_{r1} + F(\mathbf{a}_{r2} - \mathbf{a}_{r3}) \qquad (7)$$

where (a_{r1}) is either a randomly selected vector or it represents the best solution from the current population. $F \in [0,2]$ is a factor that controls the amplification of the differential variation. The mutation operator implemented in this work was:

$$\mathbf{v}_i = \mathbf{a}_{best} + F(\mathbf{a}_{r1} + \mathbf{a}_{r2} - \mathbf{a}_{r3} - \mathbf{a}_{r4})$$
$$i = 1,...,\mu \qquad (8)$$

The crossover operator increases the diversity of the mutated vector by means of the combination of two solutions:

$$a'_{ji} = \begin{cases} v_{ji} \text{ if } randb \le CR \text{ or } j = randr \\ a_{ji} \text{ if } randb > CR \text{ and } j \ne randr \end{cases}$$

$$i = 1,...,\mu; j = 1,...,q \qquad (9)$$

where v_{ji} is the j-th element of the mutated vector v_i, a_i is a so-called target vector, against which each new solution is compared to, and *randb* is a uniform random number from [0,1]. $randr \in [1,2,...,q]$ is a randomly chosen index, and *CR* is a parameter [0,1], which specifies the crossover constant. Selection is implemented only on the basis of a comparison between the cost function value of the new solution a'_i and that of the target vector a_i. This means that if $J(\mathrm{a}'_i) < J(\mathrm{a}_i)$ the new solution becomes a member of the population otherwise the old solution is retained. The inner loop in figure 1 implies that there are μ competitions at each generation since each member of the population plays the role of a target vector once.

Two extensions have been introduced in the original differential evolution algorithm. The first one is related to the fact that the controls are bounded so a clipping technique is introduced to prevent inadmissible solutions.

$$u_j = \begin{cases} u_j = \alpha_j \text{ if } u_j > \alpha_j \\ u_j = \beta_j \text{ if } u_j < \beta_j \end{cases}, j = 1,...,q \qquad (10)$$

where α and β represent the lower and upper boundary of the control variables, respectively. The second modification is due to the fact that the optimal control of nitrate in lettuce presents constraints at harvest time (final time). The solution computed is based on the use of penalty functions. The augmented performance index is given by

$$J' = J + \lambda(g) dist(\mathrm{x}(t_f)) \qquad (11)$$

where $\lambda(g)$ is either a penalty factor depending on the current generation or a constant penalty factor, J is given by equation (3), and two options for *dist* are:

$$dist(\mathrm{x}(t_f)) = \begin{cases} |\mathrm{x}_d - \mathrm{x}(t_f)| \\ [\mathrm{x}_d - \mathrm{x}(t_f)]^T[\mathrm{x}_d - \mathrm{x}(t_f)] \end{cases} \qquad (12)$$

4. NUMERICAL RESULTS

4.1. A Solution obtained by a gradient algorithm.

A solution of the optimal control of nitrate in lettuce was obtained by a classical first order gradient method. However, in contrast to the solution reported previously (Stigter and Van Straten, 2000) here the Adjustable Control Weight (ACW) gradient algorithm from Weinreb (1985) was implemented in order to deal with the constraints of the controls properly. This method uses an adjustable weighting matrix to modify the variation of the controls in the neighborhood of the hard control bounds. The ACW gradient method solves the continuous-time optimal control problem according to the Pontryagin's Minimum Principle. The required gradients were calculated analytically. The final time was specified at $t_f = 60$ days. The constraints at the final time were $y_{dfm} = 400$ grams of head fresh weight, and

$y_{dNO_3} = 3500$ ppm of nitrate concentration. The bounds on the controls were as follows: $0 \le I(t) \le 17.5$, $0.01 \le C_{Ca}(t) \le 0.04$, and $10 \le T(t) \le 30$. From several optimizations which were initialized with constant values for the controls a solution with a performance J*=306.3420 mol PAR $m^{-2}s^{-1}$ was obtained. For one of the optimizations figure 2 shows the calculated optimal trajectories of fresh matter and nitrate content obtained after 3000 iterations of the ACW gradient method. The optimal solution satisfies almost exactly the two constraints at harvest time. The error for fresh head weight was 0.0012 grams, and 0.1057 ppm in case of nitrate concentration.

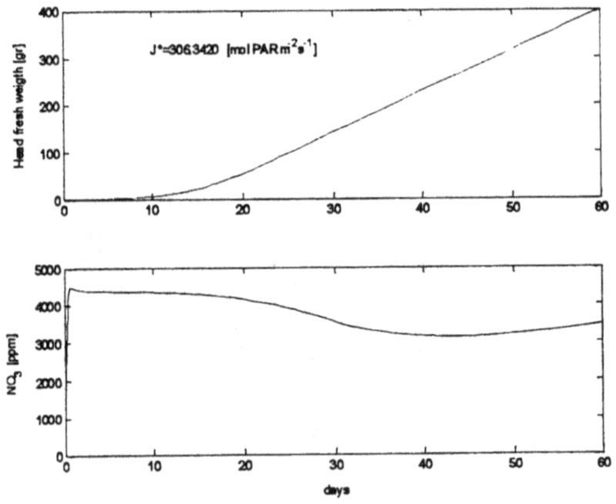

Figure 2. Optimal trajectories of head fresh weight and nitrate concentration calculated by the ACW algorithm.

Figure 3 presents the optimal control: light, carbon dioxide and temperature. Since one goal of the optimal control problem formulation was the minimization of the integral of light, the calculated optimal solution shows that it is possible to control the nitrate level at harvest time by increasing the supply of carbon dioxide and decreasing the temperature. This result confirms that with artificial light it is possible to control nitrate levels in lettuce through the control of the shoot environment (Seginer et al., 1998). The optimal trajectory of carbon dioxide supply is at the upper specified limit (0.04 mol m^{-3}), which is consistent with the fact that no cost was associated to it in the formulation of the optimal control problem.

The optimal trajectory of temperature presents a trend that goes down across the cultivation period. On the other hand, the optimal trajectory of light presents a sharp increase at earliest days, which can be explained by the demand of photosynthetic activity required to produce the desired fresh weight. Next, for the rest of the growing period, the amount of light is increased but not too much in order to meet the specification of the performance index and also to come up to the desired fresh head weight and the nitrate content at harvest time.

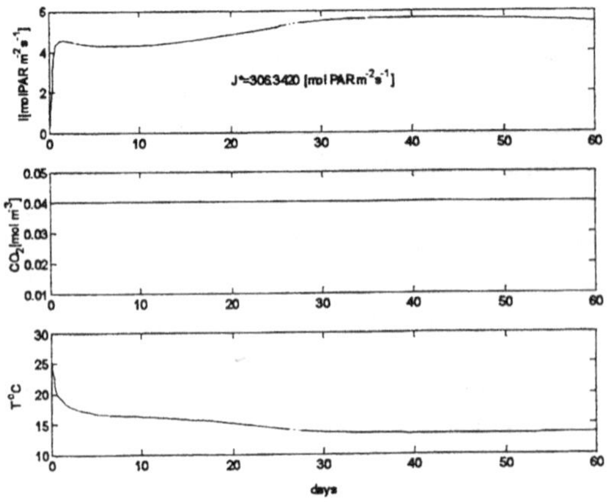

Figure 3. Optimal control inputs of light, carbon dioxide and temperature calculated by the ACW algorithm.

4.2. *A solution obtained by a Differential Evolution algorithm.*

In order to solve the previous optimal control problem by the differential evolution algorithm, the first step is the selection of a reasonable parameterization for the control inputs. For the sake of keeping the number of parameters to be optimized as small as possible, only twenty time intervals (*N=20*) were selected. Then a piece-wise constant approximation for the three controls (*m=3*) was chosen:

$$u(t) \cong u(t_k) = u_k^i \;\; t \in [t_k, t_{k+1})$$

$$k = 1, \ldots, N \, , i = 1, \ldots, m \qquad (13)$$

Therefore the optimization problem has 60 parameters. It was observed that the DE algorithm works much better solving the optimal control problem with state constraints instead of working with the original outputs head fresh weight and nitrate concentration. For that reason, using the desired values of both head fresh weight (y_{dfm}) and nitrate concentration (y_{dNO_3}), the desired values of the states at harvest time (x_d) were calculated. The next step consists of the selection of the design parameters (population size, mutation and crossover constant) in the differential evolution algorithm. Roughly speaking, using a greater population size (μ) the algorithm has more chance to convergence to a global optimum at the expense of more computation time. A population size around the dimension of the optimization problem is a good starting point but, sometimes, smaller values are enough to get good results. Greater values of the mutation constant (F) make it possible to explore the whole search space and to prevent premature convergence. Crossover constant (CR) values around one speed up the convergence of the algorithm. So, a compromise has to be established among these three parameter values in the DE algorithm. With respect to the penalty functions, several options were tested. However, better results were obtained by using varying penalty coefficients ($\lambda(g)$), which were changed exponentially according to the generation number. They were used together with the absolute difference between the desired and calculated state values for the function $dist(x(t_f)$. This approach has been applied in a similar manner by Smith and Stonier (1996) in other evolutionary algorithms.

After some experiments with several values of the population size it was observed that even with a relatively small value for the population size ($\mu = 20$), F=0.5, and CR=0.2 a good solution with a performance index of J*=317.0987 mol PAR $m^{-2}s^{-1}$ was obtained. The penalty coefficients changed exponentially from $\lambda(0) = 50$ to $\lambda(g_{\max}) = 100$. The number of generations was 5000. The deviations from the desired outputs values were 0.2130 grams for fresh weight, and 0.4272 ppm for nitrate concentration. Thus, the final constraints are almost satisfied using relatively small values for the penalty coefficients. Figure 4 presents the optimal trajectories of head fresh weight and nitrate content calculated where J*=314.6248. The deviations from the final desired outputs were 1.2418 grams and 9.9060 ppm. The number of generations was 10000. Figure 5 presents the corresponding optimal controls. The shape of the sub-optimal trajectories calculated by the DE algorithm are different from those obtained by a gradient method but clearly they show a trend that resemble the controls presented in figure 3.

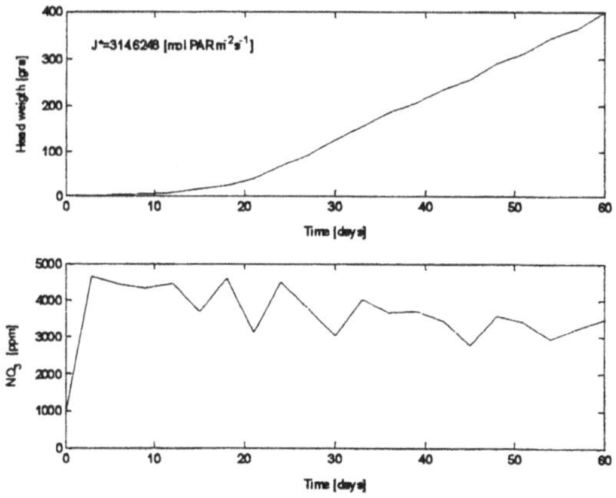

Figure 4. Near-optimal trajectories of head fresh weight and nitrate concentration calculated by a DE algorithm.

Looking at the rapid change during the first days of the optimal trajectory of light and temperature, calculated by the gradient method, it is clear that the piecewise constant control parameterization used by the DE algorithm is not able to approximate the continuous-time solution accurately. As a result the optimal performance found by the classical method is better than that found by the DE algorithm, which is only, near optimal.

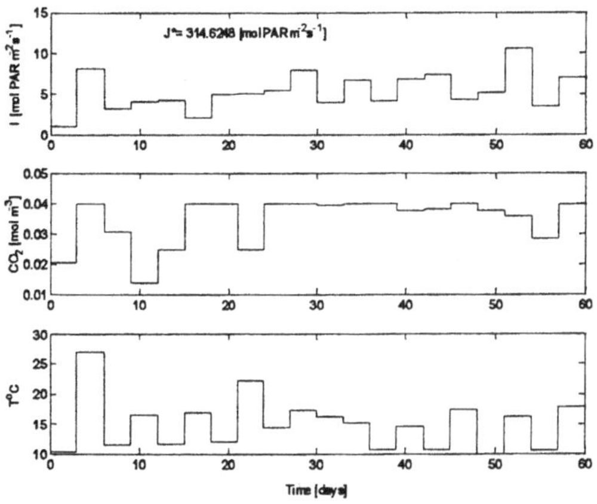

Figure 5. Near-optimal control inputs of light, carbon dioxide and temperature calculated by a DE algorithm.

On the other hand, as opposed to the classical algorithm, the differential evolution algorithm potentially finds the global solution. Therefore, efficient evolutionary algorithms as DE can be used to come up with an initial guess for classical algorithms, to prevent them from finding local minima. By increasing the number of time intervals or by specifying them as variable-length the solution of the DE algorithm will more closely approximate the continuous-time solution obtained by the classical algorithm. Finally, the rapid speed of convergence of the differential evolution algorithm near the optimal

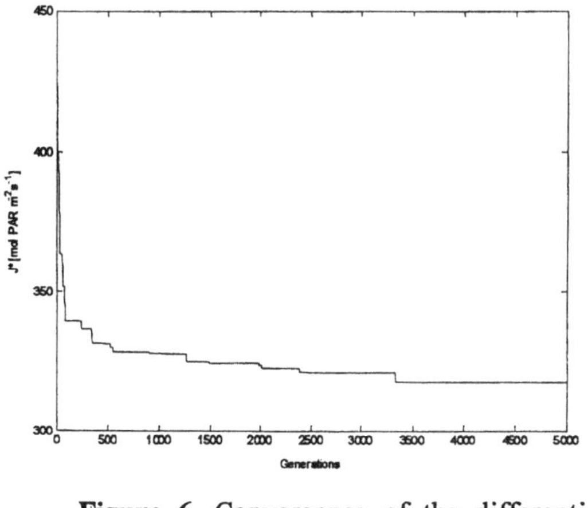

Figure 6. Convergence of the differential evolution algorithm.

solution is illustrated in figure 6. This appealing characteristic could be exploited to generate the initial guess for a local optimization method.

5. CONCLUSIONS

Through the optimal control of nitrate concentration in lettuce the benefits and drawbacks of a Differential Evolution algorithm (DE algorithm) and a classical ACW (Adjustable Control-variation Weight) gradient algorithm for optimal control were demonstrated. The Differential Evolution algorithm potentially finds the global solution whereas the classical algorithm does not. On the other hand, the classical algorithm is able to find the continuous-time solution and is more efficient, even though compared to many other evolutionary algorithms, the DE algorithm is highly efficient. Therefore, taken together, i.e. using the DE algorithm to compute an initial guess for the classical algorithm, an algorithm can be obtained that combines the advantages of both approaches.

REFERENCES

Bryson A. E. Jr., *Dynamic Optimization*, Addison Wesley, Menlo Park, 1999.

Ioslovich I, Seginer I., Acceptable nitrate concentration of greenhouse lettuce and optimal control policy for temperature plant spacing and nitrate supply, Preprints Agricontrol 2000, International conference on Modelling and control in agriculture, horticulture and post-harvested processing, July 10-12, 2000, Wageningen, The Netherlands, pp. 89-94.

Seginer, I., Buwalda, F., Van Straten G., Nitrate concentration in greenhouse lettuce: A modeling study, *Acta Horticulturae* 456: 189-197, 1998.

Smith S., Stonier R., Applying Evolution Program Techniques to Constrained Continuous Optimal Control Problems, *Proceedings of the IEEE Conference on Evolutionary-Computation*, 1996, IEEE, Piscataway, NJ, USA, pp 285-290.

Stigter, J.D., Van Straten, G. Nitrate control of leafy vegetables: a classical dynamic optimization approach, Preprints Agricontrol 2000, International conference on Modelling and control in agriculture, horticulture and post-harvested processing, July 10-12, 2000, Wageningen, The Netherlands, pp. 95-99.

Storn R. and Price K., Differential Evolution-A Simple and Efficient Heuristic for Global Optimization over Continuous Spaces, *Journal of Global Optimization* 11:341-359, 1997.

Weinreb A., Optimal control with multiple bounded inputs, PhD Thesis, Stanford University, 1985.

Yarkoni, N., McKenna, P. NICOLET Simulation model B3, Agricultural Engineering Department, Technion-Israel Institute of Technology, July, 2000.

AUTHOR INDEX

www.ingramcontent.com/pod-product-compliance
Lightning Source LLC
Chambersburg PA
CBHW081149250326
R18032300001B/R180323PG41598CBX00001B/1
* 9 7 8 0 0 8 0 4 3 5 6 3 3 *